Technology and Security in the 21st Century

A Demand-side Perspective

Stockholm International Peace Research Institute

SIPRI is an independent international institute for research into problems of peace and conflict, especially those of arms control and disarmament. It was established in 1966 to commemorate Sweden's 150 years of unbroken peace.

The Institute is financed mainly by a grant proposed by the Swedish Government and subsequently approved by the Swedish Parliament. The staff and the Governing Board are international. The Institute also has an Advisory Committee as an international consultative body.

The Governing Board is not responsible for the views expressed in the publications of the Institute.

Governing Board

Ambassador Rolf Ekéus, Chairman (Sweden)
Sir Marrack Goulding, Vice-Chairman (United Kingdom)
Dr Alexei G. Arbatov (Russia)
Dr Willem F. van Eekelen (Netherlands)
Dr Nabil Elaraby (Egypt)
Rose E. Gottemoeller (United States)
Professor Helga Haftendorn (Germany)
Professor Ronald G. Sutherland (Canada)
The Director

Director

Alyson J. K. Bailes (United Kingdom)

sipri

Stockholm International Peace Research Institute
Signalistgatan 9, SE-169 70 Solna, Sweden
Cable: SIPRI
Telephone: 46 8/655 97 00
Telefax: 46 8/655 97 33
Email: sipri@sipri.org
Internet URL: http://www.sipri.org

Technology and Security in the 21st Century
A Demand-side Perspective

SIPRI Research Report No. 20

Amitav Mallik

OXFORD UNIVERSITY PRESS
2004

327.174
m254t

OXFORD
UNIVERSITY PRESS

Great Clarendon Street, Oxford OX2 6DP
Oxford University Press is a department of the University of Oxford.
It furthers the University's objective of excellence in research, scholarship,
and education by publishing worldwide in
Oxford New York
Auckland Cape Town Dar es Salaam Hong Kong Karachi Kuala Lumpur
Madrid Melbourne Mexico City Nairobi New Delhi Shanghai Taipei Toronto
With offices in
Argentina Austria Brazil Chile Czech Republic France Greece
Guatemala Hungary Italy Japan South Korea Poland Portugal
Singapore Switzerland Thailand Turkey Ukraine Vietnam

Oxford is a registered trade mark of Oxford University Press
in the UK and in certain other countries

Published in the United States
by Oxford University Press Inc., New York

© SIPRI 2004

First published 2004

All rights reserved. No part of this publication may be reproduced,
stored in a retrieval system, or transmitted, in any form or by any means,
without the prior permission in writing of Oxford University Press,
or as expressly permitted by law, or under terms agreed with the appropriate
reprographics rights organization. Enquiries concerning reproduction
outside the scope of the above should be sent to the Rights Department,
Oxford University Press, at the address above

You must not circulate this book in any other binding or cover
and you must impose this same condition on any acquirer

British Library Cataloguing in Publication Data
Data available

Library of Congress Cataloging in Publication Data
Data available

ISBN 0-19-927175-5
ISBN 0-19-927176-3 (pbk)

3 5 7 9 10 8 6 4 2

Typeset and originated by Stockholm International Peace Research Institute
Printed in Great Britain on acid-free paper by
Biddles Ltd., King's Lynn, Norfolk

Contents

Preface	*vii*
Foreword	*ix*
Acronyms	*xi*

1. Introduction: technology and security in the 21st century

I.	Background	1
II.	Technology diffusion	4
III.	Proliferation and disarmament	7
IV.	Technology controls	14
V.	New technologies and new concerns	16
VI.	Organization of this report	19

2. Changing threat perceptions and proliferation concerns — 22

I.	The changing world order	22
II.	The regional impact of conventional arms transfers	31
III.	Future conflicts and threat perceptions	35
IV.	Causes of WMD proliferation	42
V.	Prioritizing proliferation concerns	51
VI.	Future challenges	55
Table 2.1.	Five recipients of US aid related to the war on terrorism	32
Figure 2.1.	National security parameters	24

3. A different perspective on arms control and export control regimes — 59

I.	Arms control and disarmament	59
II.	The Non-Proliferation Treaty and the Comprehensive Nuclear Test-Ban Treaty	70
III.	The changing environment for export controls	76
IV.	Missile proliferation, the Missile Technology Control Regime and ballistic missile defence	82
V.	The efficacy of multilateral export control regimes	92
Table 3.1.	Membership of multilateral weapon and technology transfer control regimes, as of 1 January 2004	60

4. Technology diffusion in the 21st century — 102
    I.   Technological interplay — 102
    II.   The impact of new technologies — 104
    III. Future trends in technology and strategy — 113
    IV. New patterns of technology diffusion — 117

5. Conclusions — 122
    I.   In summary — 122
    II.   The future of technology controls — 128
    III. A new approach to technology management — 131
    IV. Conclusions and recommendations — 135

Index — 143

Preface

During the widespread reappraisal of dangers from the proliferation of weapons of mass destruction (WMD) that has followed the strategic shock of the events of 11 September 2001, the problem of technology transfer has regained a saliency it last enjoyed at the time of the cold war. In that era, Western controls on the export of strategically sensitive items (conventional weapons and dual-use technologies as well as WMD) were directed against the single manifest threat of the Soviet Union and its allies. Today, the top concerns are about how to stop such capabilities getting into the hands of a range of states around the world that are considered unreliable and destabilizing, or—the worst nightmare of all—into the hands of terrorists. While the stakes are high, the odds are also stacked against easy or rapid success. Not only have the most dangerous technologies already been diffused so far that 'secondary' producers, in addition to the world's most developed nations, can supply them, but the conditions of globalization make it far easier for state and non-state actors alike to gain access to dangerous knowledge and materials by diverse and virtually untraceable paths. At the same time, the adequacy of established arms control and export control approaches—especially those enshrined in international legal and institutional forms—has been called in question by the world's one remaining superpower, the United States, which has preferred to tackle its most urgent related concerns by executive, coercive and (in Iraq) unilateral military action.

This Research Report is designed to contribute to the ongoing debate on policy remedies for proliferation-related technology diffusion from an often neglected, but important, point of view: that of substantial military powers in the developing world. Drawing on the experience of a distinguished career in the public service of India, Amitav Mallik points out that technology diffusion can be, from the recipient's viewpoint, the key not only to maintaining acceptable levels of national security but also to promoting broader economic and social development. He argues that traditional technology transfer controls, often prolonged too uncritically after their initial cold war rationale had departed, can mask double standards (since nuclear-weapon states continue to 'proliferate' their own arsenals) and at best tend to perpetuate an unequal division of the world's peoples. Some current trends—such as greater leniency towards perceived anti-terrorist 'allies', while supposed 'rogue' states are persecuted without regard for what their real security concerns may be—are bringing the system into greater disrepute, without doing anything to address the broader forces pressing open the flood-gates of diffusion.

Amitav Mallik makes his case from the avowed starting-point of India's own experience and interests, but he advances arguments in which many other regional powers of the developing world may find their concerns reflected. He strongly advocates the elimination of chemical and biological weapons and a step-by-step approach to a nuclear weapon-free world, beginning with the more universal acceptance of bans on testing, nuclear first use and the further production of fissile materials. For the immediate future, however, he argues that a universal buy-in to the supreme aims of arms control—and the tightening of the noose around the limited number of real enemies of international society—is likely to be best served by a regime more open to the idea of North–South technology sharing. States' suitability for such sharing, and the international acceptance or non-acceptance of their nuclear capability, should be determined less by where they lie in the world and more by their characteristics and demonstrated behaviour under a range of security-linked criteria. A design for achieving this approach, packaged with other high-priority arms control measures, is presented in the closing chapter. His proposals—but even more, the viewpoints and the reasoning that lie behind them—offer an intriguing and challenging contribution to the current debate on how to modernize (and whether to 'universalize') the world's inherited systems of arms and export control for a new age of globalization and asymmetric threat.

A book containing such novel viewpoints presents many challenges, and a long process of teamwork has gone into producing this one. My special thanks are due first and foremost to Amitav Mallik for the devoted work he has put into writing it both in India and during visits to SIPRI, but also to editors Andy Mash and Teslin Seale, and to Ian Anthony and Shannon N. Kile at SIPRI for valuable professional advice given at several stages in the research and drafting process. The views expressed in the book remain Professor Mallik's own.

Alyson J. K. Bailes
Director, SIPRI
August 2004

Foreword

Throughout history, the evolution of war-fighting strategies has depended in large part on the level of technology available to warriors and leaders. Today, while basic human instincts remain much the same, technology has multiplied the human capacity to cause damage and destruction. Nations, like human beings, compete either by raising themselves to higher levels of techno-economic performance or by keeping others down, technologically and economically.

As sovereign nations around the globe struggle to gain better positions vis-à-vis other nations—with or without open conflict—the competing forces are translated into 'threats' to the well-being or sovereignty of one nation or another. A richer and a poorer nation can each have various reasons to see the other as a threat. As is common in any human endeavour, however, the most advanced or powerful players invariably define the norms and set the standards. Most others respect and aspire to such power and often form alliances to partially achieve their aims. Among nations that cannot keep pace with the powerful, some pursue their struggle alone and embittered, while others become rebellious and seek a certain nuisance value in order to propel themselves up the ladder. In all these dynamics, whether at national or international level, it is the quality of strategy and technology that provides the vital edge. Technology has been and remains the prime instrument for enhancing security and development, while maintaining superiority also by the denial of technology to others

In the 21st century technology is advancing very rapidly, converting yesterday's fiction into today's reality. The most effective users of technology have become the advanced group of nations, while in the less privileged 'third world' some progressive developing nations have moved faster in this respect than others. Technology access and technology denial have played a major role in this division of humanity.

Technology today has become so intrinsic to human life that it is taken for granted. Its contribution is recognized more by its absence, when it is not available or denied. The pursuit of technology is thus very natural to human beings and nations: whether small or big, more developed or less developed; whether it be for survival, defence, development, or at times for more devious designs. Notwithstanding the negative implications, diffusion and proliferation of technology is unavoidable in today's interdependent world.

Real solutions for international peace and stability will have to be based on the sharing of technology benefits by all human beings without artificial discrimination. However, in the real world all cannot be equal. The future of mankind will therefore depend on the skilful management of technologies in a manner designed to reduce conflicts and to foster cooperation and peaceful coexistence. This will be the greatest challenge of the 21st century.

Professor Amitav Mallik
August 2004

Acronyms

ABL	Airborne laser
ABM	Anti-ballistic missile
ACM	Advanced Cruise Missile
AG	Australia Group
ASAT	Anti-satellite
ASEAN	Association of South-East Asian Nations
BMD	Ballistic missile defence
BTWC	Biological and Toxin Weapons Convention
BW	Biological weapon
C^3	Command, control and communications
C^3I	Command, control, communications and intelligence
CAV	Common aero vehicle
CBM	Confidence-building measure
CBW	Chemical and biological weapon
CD	Conference on Disarmament
COCOM	Coordinating Committee for Multilateral Export Controls
CTBT	Comprehensive Nuclear Test-Ban Treaty
CW	Chemical weapon
CWC	Chemical Weapons Convention
DARPA	Defense Advanced Research Projects Agency (US)
DEW	Directed energy weapons
EU	European Union
FMCT	Fissile material cut-off treaty
G7/G8	Group of Seven/Eight industrialized countries
GBI	Ground-based interceptor
GPS	Global Positioning System
HCOC	Hague Code of Conduct (Against Ballistic Missile Proliferation)
HCV	Hypersonic cruise vehicle
HEU	Highly enriched uranium
HPL	High power lasers
IAEA	International Atomic Energy Agency
ICBM	Intercontinental ballistic missile

IT	Information technology
JDAM	Joint Direct Attack Munition
LEO	Low earth orbit
MAD	Mutual assured destruction
MANPADS	Man-portable air defence systems
MEADS	Medium Extended Air Defense System
MIRV	Multiple independently targetable re-entry vehicle
MTCR	Missile Technology Control Regime
MW	Megawatt
NATO	North Atlantic Treaty Organization
NBC	Nuclear, biological and chemical (weapons)
NCW	Network-centric warfare
NFU	No first use
NNWS	Non-nuclear weapon states
NPT	Non-Proliferation Treaty
NSG	Nuclear Suppliers Group
NTM	National technical means
NWS	Nuclear weapon states
P5	Permanent members of the UN Security Council
PAC-3	Patriot Advanced Capability
PTBT	Partial Test Ban Treaty
R&D	Research and development
ROOT	Responsible Ownership of Technology
SALW	Small arms and light weapons
SBIRS	Space Based Infrared System
SDI	Strategic Defense Initiative
SLV	Space launch vehicle
THAAD	Theater High-Altitude Area Defense
UAV	Unmanned air vehicle
UN	United Nations
UNROCA	UN Register of Conventional Arms
VLSI	Very large-scale integration
WA	Wassenaar Arrangement
WMD	Weapons of mass destruction

1. Introduction

I. Background

The importance of modern technology in security and world affairs has never been so convincingly demonstrated as in the military operation in Iraq of March–May 2003. The overwhelming techno-military dominance of the US-led coalition was vividly displayed on millions of television screens across the world, albeit against an enemy without even remotely comparable military capacity. The successful application of high-technology weapons and techniques was a demonstration of how, since the end of the cold war, the sole superpower has been sharpening its superior conventional military technological strength while much of the world was wishfully looking for a peace dividend. It was with the latest in technology that the US-led coalition was able to wage a high-precision war on a distant land to achieve its initial political aims with ease.

The Iraq war demonstrated that the resolution of satellite reconnaissance imagery has been vastly improved to provide details down to a scale of centimetres and that the Global Positioning System (GPS) has been made simple, reliable and secure for ready use by the foot soldier. The command, control, communications and intelligence (C^3I) network was more effectively integrated than ever and experienced few failures in spite of adverse circumstances, bringing closer the goal of fully realizing the potential of the 'digital battlefield'. Unmanned air vehicles (UAVs) were upgraded for effective use for surveillance (following their use in combat operations in Afghanistan and in Yemen). Combat aircraft operated freely in enemy territory after disabling or destroying all enemy electronic and air-defence capabilities. AH-64 Apache helicopters were used extensively in an integrated real-time system for effective air support to ground forces that were themselves well equipped with modern technological aids and chemical protection suits. The 'network-centric' operation reduced the 'sensor to shooter' time gap dramatically to permit unprecedentedly quick reactions, almost in real time, to the battlefield situation. It is the combination of technology and strategy that has raised the United States well above the rest of the world, almost out of reach of even of the second-best.

As regards weapon technology, the advanced Block III Tomahawk missiles worked with satellite guidance from a distance of thousands of kilometres and Joint Direct Attack Munition (JDAM) precision-guided weapons were operational irrespective of weather or visibility, thanks to GPS signals and advanced guidance technologies. The BLU-114/B 'soft bomb' was available to disable electrical power grids by releasing carbon-fibre filaments. The CBU-97 smart bomb used passive infrared sensors to locate and hit enemy targets with high precision and reliability. A variant of the same bomb—the CBU-105—is capable of sprouting 40 precision heat-seeking bullets from a mother bomb dropped from a height of 12 kilometres. Ten sub-munitions individually float towards their targets to eject four hockey puck-size projectiles with infrared seekers to home in on targets such as tanks or other armoured vehicles and fire armour-piercing pellets to neutralize the target.[1] These are just some of the many advanced weapons which were used in Iraq.

Seen with clinical detachment as to its causes and effects or its future implications, the Iraq war was conventional defence technology on display at its best. What was not so visible included the vast network of computers and sophisticated software that made it possible to integrate all the elements of this huge war-fighting system from a remote centre with full flexibility and constant control. Space-based surveillance, convincing air superiority, Advanced Cruise Missile (ACM) technology, precision-guided weapons, combat UAVs and the emerging range of new-technology weapons have changed the ground realities of conventional war. Stretched to the limits of asymmetry, as in the case of the Iraq operations, this technological superiority gave its possessor the option to use military force, in preference to diplomatic means, to enforce a change of regime in a distant sovereign country. In this case, there was overwhelming international agreement on the end objective but little agreement on the method used. However, the power of technology had the last say in the matter, setting a new precedent for the 21st century.

The operation was a first of its kind in many different ways and thus will be under intense analysis in many forums across the world: whether in terms of its impact on the efficacy of United Nations (UN)

---

[1] For a convenient index of technical descriptions of US weapon systems see Federation of American Scientists Military Analysis Network, 'United States weapon systems', URL <http://fas.org/man/dod-101/sys/index.html>.

systems or the future of multilateral approaches to arms control. For defence analysts, it has undoubtedly set new benchmarks for strategies and techniques for winning future wars or for averting future attacks. What this convincingly highlights is the decisive role of technology in matters of 'defence preparedness and national security' for all nations of the world, whether developed or developing, big or small, democratic or autocratic. While this lesson has clear potential to exacerbate the race for dual-use high technology for defence and development, it might also be hoped that the 'shock and awe' power of high-tech conventional weapons could set the stage for a reduced dependence on weapons characterized as weapons of mass destruction (WMD). However, continued dependence on, and new uses for, nuclear weapons—as brought out in the 2001 US Nuclear Posture Review—has thus far preserved or even strengthened their value as the ultimate deterrent.[2] This will certainly make disarmament and non-proliferation objectives much more difficult to achieve in the foreseeable future.

The 20th century witnessed a phenomenal growth in technology, for both military and civilian applications. During World War II it was the military strategists who called for superior technology to remain ahead of the enemy. During the cold war it was the race for superiority in high technology that pushed WMD capabilities and space-exploration technologies. A 2001 RAND report candidly accepts that 'almost everything we've ever done in space has been predominantly motivated by the security perspective'.[3] Whether it is the miniaturization in electronics achieved through very large-scale integration (VLSI) techniques or high-sensitivity video camera technology, most high-tech developments in the past few decades have been driven by the cold war's security imperative to maintain the technological edge.

Organizations such as the US Defense Advanced Research Projects Agency (DARPA) have opened up new frontiers in science and technology that promise the almost impossible, converting yesterday's science fiction into today's reality. Armed with the confidence of high-tech aids, military strategists are already speaking of 'total

---

[2] The Nuclear Posture Review is a classified report. A briefing on the public aspects is available in US Department of Defense, 'Special briefing on the Nuclear Posture Review', 9 Jan. 2002, URL <http://www.defenselink.mil/news/Jan2002/t01092002_t0109npr.html>.

[3] Preston, B. *et al.*, *Space Weapons, Earth Wars*, RAND Report MR-1209-AF (RAND Corporation: Santa Monica, Calif., 2001), URL <http://www.rand.org/publications/MR/MR1209>.

information awareness', and yet the issues of information dependence and information security are facing new challenges from the technology being used. Increasing computing capacities and shrinking hardware sizes are creating a new environment where the developed nations are losing the comfort zone of the technological edge. Instant worldwide communication, instant media coverage and the compelling market forces of modern economics are all bittersweet fruits of the technology revolution that are affecting quality of life as well as security perceptions. The impact of technology will be even more pronounced in the future, and hence the technology control mechanisms of the future will have to face new challenges, some of which are yet to be identified clearly.

While the military operation in Iraq will remain a major milestone of the 21st century—and one where the USA was in the driver's seat—the other equally unforgettable event of this young century which occurred on 11 September 2001 saw the USA under attack in its own homeland. Terrorist networks have learned how to exploit high-tech mechanisms such as the Internet and global banking systems to raise funds, plan, coordinate and communicate. The menace is no longer confined to small, disgruntled religious fanatic groups. Extremism has been deliberately fuelled and funded by short-sighted or dictatorial state actors that want to use these elements for their military and diplomatic advantage. Past analysis has shown that terrorism could not have gained this amorphous global presence without state-level support.[4] Given the effects of globalization and the inevitable diffusion of technology, protecting sensitive technologies from violent misuse will be the major and immediate challenge for arms control during the early decades of this century.

II. Technology diffusion

The processes of the development and dissemination of technology through increased international trade and interdependence have evolved significantly since the end of the cold war. The drivers for political alignments have undergone major changes and, in many developed nations, defence budgets have declined. This has created

---

[4] On state-sponsored terrorism see Laqueur, W., *The New Terrorism: Fanaticism and the Arms of Mass Destruction* (Oxford University Press: New York, 1999), pp. 156–83; and Schmid, A. P. et al., *Political Terrorism: A New Guide to Actors, Authors, Concepts, Data Bases, Theories and Literature* (North-Holland: Amsterdam, 1988).

new pressures in technologically advanced countries to increase arms exports or to apply the defence technology base to civilian markets in order to generate wealth from the technology they have developed. While there is increasing agreement about the need for export controls to prevent the most dangerous technologies going to dangerous actors, some diffusion of militarily relevant technology becomes inevitable in the remaining broad field of technology trade. The grey zone of potential but undetermined risk, where international agreements are difficult to achieve, is likely to become broader with the continuous evolution of technology.

There is now more interplay between civilian technologies and military technologies. Defence research was the driver for technological growth during the cold war years and there were many spin-off benefits for the civilian sector. Dual-use technology was defined as the class of military technologies that also had clear civilian applications. However, in the past few decades it has increasingly been the civilian sector's research and development (R&D) that has provided technology options for military applications. Dual-use technology is becoming defined as civilian technology that could also be diverted for military applications. Controlling or denying technology to potential technology developers or even recipients will be more difficult in the light of the commercial value of such technologies and also because of the greater role of market forces and the declining influence of government controls on technology transfer interactions. Apart from direct arms sales or military assistance, most major technology transactions now take place within and between multinational companies that are spread worldwide.

Another feature of high-tech commerce is that, as technological sophistication increases, the cost of development also rises sharply, forcing industries to depend heavily on export markets. In such a buyer's market, the buyer is no longer satisfied with the product alone and some form of technology transfer connected to the manufacturing or production processes is almost invariably part of the contract.[5] This assists the technology diffusion process and can create secondary suppliers that are capable of further innovation and that can offer alternative products in areas where a leading supplier from the industrial-

---

[5] Willett, S. and Anthony, I., *Countertrade and Offsets Policies and Practices in the Arms Trade*, Copenhagen Peace Research Institute Working Paper no. 20 (Copenhagen Peace Research Institute: Copenhagen, 1998).

ized group of nations may not wish to compete for political or economic reasons. Larger companies are more amenable to compliance with export controls because their large scale of operations and greater staying power provide them with greater margins to absorb short-term losses. Smaller companies that are dependent on innovation may represent more of a challenge to the implementation of export controls.

Another factor worth noting is the different types of exchange dynamic for different levels of technology. For example, controls on fast and high-capacity computers were a major focus during the 1970s and 1980s. However, most commercially available computers are now orders of magnitude more efficient than those subjected to tight control just 20 years ago. In addition, the power of software today is so much greater that it allows highly innovative uses of ordinary computing hardware. The same is true, in the communications field, of sensor technology or even material technology. The technologies of most potential importance for the future will be information technology (IT) and systems technology. In both these areas, human ingenuity will be more important than components or sub-systems. Technology exchanges between suppliers and recipients will therefore be of a different nature, where both may stand to gain equally in the final analysis. This is the true driving force for modern technology diffusion.

Thanks to technology diffusion, the world today is far more interdependent and interactive. This has created an unprecedented acknowledgement of the need to work together for enhanced stability and peace in the world, with a focus on diminishing violence and instability. On the other hand, increased information may feed suspicions and perceptions of threats, while the instant media coverage of the most newsworthy events can create sharper reactions and additional pressures for action. Overall, public awareness has increased—for better or for worse.

A particular impact of military technology advances is the effective enlargement of the so-called battlefield. The traditional focus on the well-defined front line of battle has been altered by technological capabilities that can reach deep into enemy territory, thereby making additional parties vulnerable to the overspill from a particular theatre of battle. Since modern long-range weapons and new 'weapons of mass disruption'[6] can expand the scale of conflicts swiftly, the international community will be required to intervene quickly lest regional conflicts spill over to other regions and take on global dimensions. Technology also enhances the tempo of operations and reduces the reaction time for decision making.

These are the technological realities of today that have significant bearings on the security perceptions and defence operations of individual nations. How individual nations will adapt to these new technology options is as yet unclear. What is clear is that, with the new level of awareness of the advantages of established technologies, the pursuit of 'high-value technology' is going to be ever more intense. A broader distribution of these technologies will become necessary if nations are to be prepared to cooperate to conduct international affairs in a more harmonious manner for future global peace and stability. The denial of legitimate requirements for high-value technology will increase tensions and create avoidable animosity.

III. Proliferation and disarmament

The political and economic imperatives that impact on the diffusion of technology relevant to non-proliferation and export controls have changed so much since the end of the cold war that those responsible for implementing controls are finding it difficult to achieve consensus on what to limit, to whom and by how much. The real problem for the future will be how to manage technology to ensure universal economic growth and development, while also protecting technologies from misuse so that regional and global security and stability concerns are not compromised. This kind of balanced management of technology will demand the cooperation of both the technology provider and the technology user. The mere extension of export control regimes along the lines that have been followed in recent years may no longer be viable because, as explained above, technology ownership has broadened enormously and the demarcation line between the technology provider and user is becoming increasingly diffuse.

[6] Zanders, J. P., 'Weapons of mass disruption?', *SIPRI Yearbook 2003: Armaments, Disarmament and International Security* (Oxford University Press: Oxford, 2003), pp. 683–90. Zanders classifies nuclear, biological and chemical weapons as 'weapons of mass disruption', especially when used by terrorists whose main goal is not to kill but to terrorize. The term was used several years earlier to describe IT-related warfare. See, e.g., Anderson, K., 'Cyberterrorists wield weapons of mass disruption', BBC News Online, 22 Feb. 2000, URL <http://news.bbc.co.uk/1/hi/sci/tech/specials/washington_2000/648429.stm>.

8   TECHNOLOGY AND SECURITY

The genesis of preoccupations with nuclear proliferation can be traced back to the early cold war period, when it was thought necessary to prevent the revolutionary advantage of nuclear weapon capability from getting into too many hands. This was primarily because of the shocking potential of nuclear weapons for annihilating all humanity. At the same time, there was also a perception that the small number of nuclear weapon states (NWS) did not want to see their superior status diluted. However, the real proliferation of nuclear weapons during the cold war, despite the commitments of the parties to the 1968 Treaty on the Non-Proliferation of Nuclear Weapons (Non-Proliferation Treaty, NPT), was the unimaginably steep rise in the quantity and quality of nuclear weapons—primarily held by the two superpowers. The world was held hostage by the doctrine of mutual assured destruction (MAD),[7] and it would be illogical to expect that the security perceptions of other sovereign countries would remain unaffected by such a massive vertical proliferation of technology.

Although nuclear proliferation was defined in the NPT in terms of both horizontal and vertical proliferation, various mechanisms were put in place to control only horizontal proliferation. The International Atomic Energy Agency (IAEA)[8] and the Nuclear Suppliers Group (NSG)[9] were established to enforce tight controls on nuclear technology transfers and to discourage even indigenous nuclear technology development, except under close supervision. The NPT, the IAEA and the NSG have certainly succeeded in limiting the number of NWS. However, they proved unable to contain nuclear technology transfers that contributed to weapon programmes based on the political priorities of strong NWS, such as those believed to have taken place from

[7] The nuclear stalemate in Soviet–US relations in the 1960s led to the adoption of the doctrine of mutual assured destruction, according to which no country would attack another if it knew that the attacked side had the capability to inflict unacceptable damage on the attacker.

[8] The IAEA is an intergovernmental organization within the UN system. The IAEA is endowed by its Statute, which entered into force in 1957, to promote the peaceful uses of atomic energy and ensure that nuclear activities are not used to further any military purpose. Under the NPT and the nuclear weapon-free zone treaties, non-nuclear weapon states must accept IAEA nuclear safeguards to demonstrate the fulfilment of their obligation not to manufacture nuclear weapons.

[9] The NSG, established in 1975, coordinates multilateral export controls on nuclear materials. In 1977 it agreed the Guidelines for Nuclear Transfers (London Guidelines, revised in 2000), which contain a 'trigger list' of materials that should trigger IAEA safeguards when exported for peaceful purposes to any non-nuclear weapon state. In 1992 the NSG agreed the Guidelines for Transfers of Nuclear-Related Dual-Use Equipment, Material, Software and Related Technology (Warsaw Guidelines, revised in 2000). For the participants see table 3.1.

France and the USA to Israel or, more recently, from China to Pakistan.[10] The non-proliferation mechanisms certainly slowed indigenous development in India but could neither contain it completely nor reverse it. Ultimately, India, Israel and Pakistan have emerged as additional de facto nuclear weapon states—despite the indefinite extension of the NPT in 1995 in its original format of 1968.

In the changed circumstances of today, the perceived problem of the proliferation of technology extends well beyond nuclear missile-related issues and includes other types of WMD, as well as potentially powerful conventional weapon technologies that are too diverse to control and are within the reach of many nations. Hence, the non-proliferation focus is now shifting to certain countries of concern and is increasingly based on an assessment of their inclinations or intentions to use their technological capabilities, depending on their past record of good or bad behaviour. The focus is thus changing from technology to 'technology user'. This is primarily because a certain degree of technology diffusion is now accepted as inevitable and the trend for this diffusion to increase is being recognized.

The fundamental challenge for the future of non-proliferation and arms control will be how to balance competing interests among various nations on the issues arising out of technology diffusion. On the one hand, it is imperative for regional and global security that the misuse of technology should be blocked across the board while, on the other hand, it is equally imperative that healthy and stable international trade is encouraged to keep pace with overall socio-economic development and constantly evolving advances in technology. It is important to understand that technology is not merely a product or an artefact but represents the application of technical knowledge to serve a given purpose with greater efficiency or ease, with the help of technical options. Often the desired application is identified first and the technological innovation to realize it follows later. A mechanism is needed to define the (probably narrow) areas of technology that will be closely controlled. A complementary mechanism is also needed to target the control of a much broader band of technologies on particular end uses and end users. This is in fact the current trend among those states that cooperate to develop and enforce national export controls. However, to be sustainable such a system must shift to consider

---

[10] The development of the nuclear arsenals of Israel and Pakistan is discussed in section IV of chapter 2 in this volume.

how best to influence the decisions of those for whom technology options may be available independent of the actions of the relatively small group of participants in the existing control regimes.

Space technology and missiles

There was also a widespread proliferation of space launch technology during the cold war. It was triggered by the Soviet launch of the Sputnik space capsule in 1956, ahead of US space technology efforts. This energized the USA to race ahead to send a man to the moon before the Soviet Union could do so, and the ensuing space launch technology race produced the massive technology infrastructure for the evolution of missile technology. By the 1950s, missiles were seen to be extremely attractive for the delivery of strategic or tactical weapons to remote enemy territories. The initial proliferation in terms of numbers as well as technological sophistication was, once again, initiated by the two superpowers. This set the stage for a new race for missile capabilities by several countries that were pursuing the same technology for space applications. Although 'cooperation for the peaceful use of space'[11] was the theme during the initial years of space exploration, even civilian scientific uses were undertaken in large part for security reasons in the context of the cold war, as confirmed by the RAND report referred to above.[12]

The Missile Technology Control Regime (MTCR)[13] was announced in 1987 by the Group of Seven (G7) industrialized nations to prevent further proliferation of missiles that could deliver nuclear weapons.[14] The MTCR participating states have made a political commitment to

---

[11] The 1967 Treaty on Principles Governing the Activities of States in the Exploration and Use of Outer Space, including the Moon and Other Celestial Bodies (Outer Space Treaty), which entered into force in Oct. 1967, provides the basic framework of international space law. See URL <http://www.oosa.unvienna.org/SpaceLaw/outerspt.htm>.

[12] Preston *et al.* (note 3).

[13] The MTCR is an informal military-related export control regime that produced the Guidelines for Sensitive Missile-Relevant Transfers (1987, later revised). Its goal is to limit the spread of WMD by controlling both ballistic missiles and UAVs (including cruise missiles) that can be used to deliver them. For information about the MTCR see URL <http://www.mtcr.info/english/index.html>. For the participants see table 3.1 in this volume.

[14] The G8 is an informal group in which Canada, France, Germany, Italy, Japan, Russia, the UK and the USA as well as the European Union (EU) participate. The EU is represented by the President of the European Commission and by the leader of the country that holds the Presidency of the Council of the European Union at the time of the G8 summit meeting. The G7 became the G8 at its Birmingham Summit in 1998. Russia will complete the process of becoming a full member at the 2006 Moscow Summit, when it assumes the G8 Presidency for the first time.

one another to apply national export controls to items included in the Equipment, Software and Technology Annex, which is divided into two categories.[15] The MTCR participating states agreed a set of guidelines to be applied when making national export licensing decisions about any items on the control lists. In the case of Category I items, which include rocket systems (including ballistic missiles, space launch vehicles and sounding rockets) and UAV systems (including cruise missile systems, target and reconnaissance drones) with capabilities exceeding a 300-km/500-kg range/payload threshold, participating states have agreed to exercise particular caution when considering licence applications and to undertake their assessment with a strong presumption of denial.

Participation in the MTCR does not entitle states to obtain technology from another partner and there is no obligation to supply it to a partner. Nevertheless, the MTCR has been seen as a supply cartel—arguably an 'arms control' initiative in its most discriminatory form—because trade in controlled items is more prevalent among participating states while non-participants have been denied missiles that they see as necessary from a strategic perspective. The MTCR participating states argue that the patterns of approval and denial reflect the different levels of concern about nuclear, biological and chemical (NBC) weapon programmes that apply to different countries seeking access to controlled items. However, the appearance of discrimination on political rather than objective grounds is strengthened in the eyes of many states by missile technology transfers from MTCR participating states to some countries which, even if they implement the MTCR Guidelines (which are public documents available to any state),[16] are nonetheless both outside the regime and sensitive destinations in the MTCR's own terms.

As regards the focus of controls, the general perception of the demand side is that the MTCR's 'red-band' of Category I missiles, with a 300-km range and 500-kg payload qualification, fits well with the threat perceptions of the industrial supplier nations but interferes

---

[15] The legal control over exports of missiles by MTCR participating states rests on national laws and regulations. However, while missiles are controlled under national laws, licensing authorities need only apply the MTCR Guidelines to that sub-set of items included in the MTCR Equipment, Software and Technology Annex. (Category II items are those in the Annex which are not designated as Category I items.)

[16] The MTCR documents are available on the SIPRI Internet site at URL <http://projects.sipri.se/expcon/mtcr_documents.html> and at URL <http://www.mtcr.info>.

seriously with many other parts of the world where 300 km could be an adequate range for targets seen as causing vital security threats. For example, between Israel and the Palestinians, Iran and Iraq, India and Pakistan, and North Korea and South Korea there are unresolved differences that over the years have led to regional tensions and the creation of additional military capabilities, particularly missile and nuclear weapon capabilities, despite all efforts for control.

The MTCR, which is an informal agreement among a group of nations with common interests, was not conceived to achieve zero levels of missile technology proliferation. The result is that over 30 nations are believed to have varying levels of missile capability, with more countries still attempting to acquire Category II controlled items in particular. The nature of missile technology, and its commonality with space-launch technology, allows full exploitation of the narrow limits of the 300-km/500-kg threshold. In recognition of this, the MTCR Guidelines were modified in 1993, when participating states agreed to apply particular restraint (including the strong presumption of denial) when assessing any application to export any missiles, whether or not they are listed in the Annex, if it is judged that they are intended to be used for the delivery of WMD. However, the effort to make controls more comprehensive by considering end-use has in practice led to more ambiguity and subjectivity and has opened up new discriminatory imbalances. The negotiations leading to a new international code of conduct against ballistic missile proliferation, the Hague Code of Conduct (HCOC),[17] were a move in the right direction because the code calls for international consensus on the use of missile technology, rather than simply trying to control the export of equipment.

The Biological and Toxin Weapons Convention (BTWC)[18] was opened for signature on 10 April 1972 and entered into force on

---

[17] The International Code of Conduct Against Ballistic Missile Proliferation (ICOC), now known as the Hague Code of Conduct, was concluded in Nov. 2002 between a group of states concerned with preventing and curbing the proliferation of ballistic missile systems capable of delivering WMD and to strengthen multilateral disarmament and non-proliferation mechanisms. By Dec. 2003, 110 states had subscribed to the HCOC. See URL <http://www.minbuza.nl/default.asp?CMS_ITEM=MBZ460871>; and Ahlström, C., 'Non-proliferation of ballistic missiles: the 2002 Code of Conduct', *SIPRI Yearbook 2003* (note 6), pp. 749–59.

[18] The 1972 Convention on the Prohibition of the Development, Production and Stockpiling of Bacteriological (Biological) and Toxin Weapons and their Destruction prohibits the development, production, stockpiling or acquisition by other means or retention of microbial or other biological agents, or toxins whatever their origin or method of production, of types and in quantities that have no justification of prophylactic, protective or other peaceful pur-

26 March 1975. It is one of the most non-discriminatory treaties, with 163 parties and 18 signatories. The Australia Group (AG)[19] was established in 1985 to apply export controls for the serious purpose of eliminating chemical and biological weapons (CBW). The Chemical Weapons Convention (CWC), with 145 parties and 34 signatories, was concluded in 1992 but entered into force only on 29 April 1997.[20]

Despite a broad consensus on banning and eliminating these weapons, 30 years of history bear testimony to the problems of implementing such multilateral control regimes. The recent reluctance of the USA to finalize agreement on an international protocol to the BTWC and the limited success of UN inspections in Iraq, particularly in the area of CBW, leading to unilateral pre-emptive action by a US-led coalition, have seriously compromised such multilateral disarmament efforts and raised a host of new questions about treaty compliance, monitoring and verification procedures and so on.[21] These are complex and serious issues that demand a clear understanding of the technological and politico-military interconnections involved.

This report makes a broad assessment of the various proliferation issues and control regimes in the context of recent world events and trends set in the 20th century. It argues that treating different technologies for NBC weapons under one category of WMD is technically flawed because the complexities, infrastructure requirements and possible monitoring mechanisms for verification of treaty compliance are

poses, as well as weapons, equipment or means of delivery designed to use such agents or toxins for hostile purposes or in armed conflict. The destruction of the agents, toxins, weapons, equipment and means of delivery in the possession of the parties, or their diversion to peaceful purposes, should be effected not later than 9 months after the entry into force of the convention. For recent BTWC developments, including the continuing obstacles to agreement on verification provisions, see Hart, J., Kuhlau, F. and Simon, J., 'Chemical and biological weapon developments and arms control', *SIPRI Yearbook 2003* (note 6), pp. 645–82. Complete lists of parties, signatory and non-signatory states are available on the SIPRI CBW Project Internet site at URL <http://projects.sipri.se/cbw/docs/bw-btwc-mainpage.html>.

[19] The Australia Group is a group of states that meets informally each year to monitor the proliferation of chemical and biological products and to discuss CBW-related items which are subject to national regulatory measures. For the participants see table 3.1 in this volume.

[20] The 1997 Convention on the Prohibition of the Development, Production, Stockpiling and Use of Chemical Weapons and on their Destruction (Chemical Weapons Convention, CWC) prohibits the use, development, production, acquisition, transfer and stockpiling of chemical weapons. Each party undertakes to destroy its chemical weapons and production facilities within 10 years of the entry into force of the treaty. Complete lists of parties, signatory and non-signatory states are available on the SIPRI CBW Project Internet site at URL <http://projects.sipri.se/cbw/docs/cw-cwc-mainpage.html>.

[21] See Zanders, J. P. et al., *Non-Compliance with the Chemical Weapons Convention: Lessons from and for Iraq*, SIPRI Policy Paper no. 5 (SIPRI: Stockholm, Oct. 2003), available at URL <http://editors.sipri.se/recpubs.html>.

different in each case. The relevant problems of principle and practice are illustrated in particular from the viewpoint of developing nations, with a view to broadening understanding of the range of motives and interests involved.

Given the interdependent nature of the world, non-proliferation and disarmament must go hand in hand in order to achieve real and lasting gains. This report argues for the creation of a new world order where the use of WMD would be universally unviable and their use or threat of use totally unacceptable across the board. If conceived properly as an integrated techno-economic route to universal peace and stability, such an approach might have the potential to curb further undesirable proliferation or misuse of technology and might also make the goal of eventual universal nuclear disarmament realizable in the future.

IV. Technology controls

The relevance of technology to security is manifold and the role of 'technology controls' as an important tool for national security strategies is significant. The control of technology is exercised through either 'arms control' agreements or 'export control' regimes. The arms control initiatives of recent decades have proved fairly successful in reducing the proliferation of potentially dangerous technologies and weapons. However, the export control regimes have not prevented the creation of additional imbalances and may have accentuated certain regional conflicts, thereby adding to the causes of proliferation. The impact of some of the new technologies on international security and stability will be even more pronounced in the future. Changing perceptions of national security and the new dimensions of global technology diffusion will demand a fresh approach to arms control in general and export controls in particular.

The four broad objectives of arms control are: (*a*) to manage the techno-economic balance; (*b*) to reduce the possibility of war; (*c*) to reduce the consequences of war, if it happens; and (*d*) to optimize resources for defence so that economic development does not suffer unduly. During the cold war years, the Coordinating Committee for Multilateral Export Controls (COCOM)[22] was a technology embargo

---

[22] COCOM was established in 1949 by the USA and its allies to deny militarily useful technologies to the Soviet bloc countries through coordinated export controls. The participants were Australia, Belgium, Canada, Denmark, France, Germany, Greece, Italy,

INTRODUCTION 15

regime to prevent the transfer of dual-use technology and equipment to communist bloc states in the belief that such equipment and technology, if diverted to military use, could have contributed significantly to the military potential of the adversary.

In June 1992 the 17 countries participating in COCOM decided to establish a Cooperation Forum to define a successor regime for future technology controls. The Cooperation Forum, which did not immediately replace but at first existed alongside COCOM, had four objectives. These were: (*a*) to significantly ease access by East European countries to advanced goods and technology; (*b*) to establish procedures to ensure against diversion of these sensitive items to military or other unauthorized users; (*c*) to assist the East European states to develop their own export control systems; and (*d*) to provide a mechanism for further cooperation on export control matters.[23]

A decision in principle to abolish COCOM was taken in 1993. However, as noted above, it continued to exist alongside the Cooperation Forum for the next three years, while an alternative arrangement was under discussion. During this period the number of items on the COCOM control list was progressively reduced (and these items were no longer subject to embargo) and by 1996 several of the countries that had been the targets of the embargo were friends and important trading partners. There was active discussion at that time of enlarging the North Atlantic Treaty Organization (NATO) to include some of them as members. In 1996 the Wassenaar Arrangement (WA) emerged as an informal arrangement of states.[24]

The objectives of multilateral export control regimes typically include commitments: (*a*) to regulate sensitive technology transfers with potential military applications; (*b*) to introduce a licensing mechanism to institutionalize export controls; (*c*) to create a database for mutual information sharing for better coordination of controls; and

---

Japan, Luxembourg, Netherlands, Norway, Portugal, Spain, Turkey, the United Kingdom and the United States. See sections I and IV of chapter 3 in this volume.

[23] The establishment of the COCOM Cooperation Forum was announced in US Department of State, Daily Briefing 87, 2 June 1992, available at URL <http://dosfan.lib.uic.edu/ERC/briefing/daily_briefings/1992/9206/087.html>.

[24] The Wassenaar Arrangement on Export Controls for Conventional Arms and Dual-Use Goods and Technologies was formally established in 1996. It aims to prevent the acquisition of armaments and sensitive dual-use goods and technologies for military uses by states whose behaviour is a cause for concern to the member states. See also section V of chapter 3 in this volume. For the participants see table 3.1 in this volume.

(*d*) to identify countries of concern and prevent the proliferation of dual-use technologies to them.

The Wassenaar Arrangement lacks some of these features. Despite its lists of sensitive items organized under the 'munitions' and 'dual use' subheadings, the WA is, arguably, a loose technology control regime with many subjective elements and ambiguities. It is important to recognize that the WA evolved not from well-defined non-proliferation objectives, but more from the general desire for continued use of the substantial control infrastructure created during the cold war. Given current trends in technology and international affairs, such multilateral export control regimes need to be subjected to serious and realistic reforms, consistent with changing times and changing technology dimensions.

It is perhaps understandable that such reforms could not have been made during the 1990s, when the world was going through unprecedented changes. However, as the world organizes its approach for the new century, the time is ripe to consolidate understandings on these important issues and to take a radically fresh look at the various options for technology management in the years to come.

This report therefore examines the impact of technology controls on security perceptions, and analyses the complex issues of technology denial to prevent the misuse of technology and of technology availability for legitimate economic development, defence and security needs. It is natural for developing nations to follow routes in pursuit of technology for economic development (and the rapid enhancement of military technology capabilities) similar to those which today's developed nations followed during their early economic development. In order to achieve cooperative security among mature and progressive nations, technology must play its rightful levelling role in ensuring security and development for all partners. Achieving the correct balance between worldwide technology transfer and trade, and security concerns will require innovative approaches and careful planning by the arms control community. This report offers an approach to 'arms control' in the future.

V. New technologies and new concerns

The last decades of the 20th century were witness to phenomenal advances in military technologies. The post-1945 superpower competition drove a technology race that produced amazing results in a short time span. While the doctrine of MAD now seems irrelevant, new threats have surfaced *inter alia* because of the worldwide spread of religious fundamentalism and terrorism. The unavoidable spread of advanced technology has led to new and grave concerns regarding the proliferation of WMD technologies into irresponsible or unstable hands. The rapid progress in IT during the past decade alone has opened up several new possibilities for using this technology for strategic or operational advantages. Increasing computing speeds, smaller hardware and innovative software approaches are creating even more options. IT has already revolutionized the battlefield with the trend for network-centric command and control philosophies.

Another major new technological trend will be the increasing use of outer space for defence and security. In the past, most of the new and exotic technology developments in this field were oriented towards providing protection from ballistic missiles, and the announcement in 1983 of the Strategic Defense Initiative (SDI) by US President Ronald Reagan was one of the major drivers for new technologies that envisaged the military use of outer space.[25] Twenty years later, many of those fiction-like technologies are close to realization and the impact of these technologies on future security strategies is likely to be profound and long-lasting. Space is already being used for military purposes, mostly for surveillance and communication technology support, and these activities have had a significant impact on defence and security perceptions. The expected introduction of directed energy weapons (DEW) and possible increase in the military exploitation of satellite systems for combat purposes will revolutionize the future trial of strength between powerful nations (see chapter 4 in this volume).

The latest trends in technology development for military purposes also indicate a move towards miniaturization, improved efficiency and greater fungibility. As weapons become smaller and more efficient, deployment strategies and operational scenarios become more flexible, making a larger variety of options available to the user. The shrinking size and weight of strategic warheads are a classic example of how technology has made the attacker's job easier and the defender's job more difficult—creating more demand for newer technology options to meet the new level of threat.

---

[25] Jasani, B., 'The military use of outer space', *SIPRI Yearbook 1984: World Armaments and Disarmament* (Taylor & Francis: London, 1984), p. 352.

It is also interesting that, with the increasing accuracy of weapon-delivery technology and the increasing lethality of new warheads, there is growing interest in the development of non-nuclear strategic weapons. If compact, efficient conventional warheads could be delivered over intercontinental distances to achieve a reliable strike with surgical precision on strategic enemy targets, such a capability could well transform the basic concepts of strategic security.

While such new technologies may actually help the process of reducing the heavy reliance on nuclear weapons in the future, they would create enormous problems for the monitoring of conventional strategic capabilities from the traditional arms control point of view. For the sake of better transparency and more reliable information on arms transfers, the focus of future arms control or technology control will thus have to shift to a cooperative technology-management model involving all the responsible technology owners in the world. Technology may force the present generation to re-evaluate the definition of proliferation of items, and to countries, 'of concern' and should promote a keener understanding of the viability and benefits of universal nuclear disarmament, achieved in a well-coordinated, step-by-step fashion.

The worldwide spread of terrorism and religious fundamentalism has added a serious twist to security and vulnerability in modern society. These extremist elements act as criminals with no accountability to any national or international norms. As explained above, some states have misguidedly aided them with funds, arms, training and so on, but advances in technology have also contributed to the efficiency of these terrorists and their sponsors.

Another twist of technology is that some states, considering themselves protected by the deterrent effects of WMD, may feel emboldened to clandestinely support terrorism across borders to settle regional conflicts. It is these alarming trends and the potential for misuse of technology that must draw the immediate attention of the international arms control community, before they lead to new trends in asymmetric low-intensity warfare that could upset international security in unpredictable ways.

VI. Organization of this report

The period since the end of the cold war has witnessed milestone events of global impact that have created major reverberations around the world and changed the fundamental perspectives of most nations on national security. The genesis of these international events, however, can be traced back to the phenomenal advances over the past 50 years in high-value technologies that have been the cornerstone of defence and development. The interplay of technology and security has become more complex and this has changed the way we have viewed the world for over half a century. Diffusion of technology has become unavoidable in this new IT age, creating new challenges for the future of technology control. Global technology management must foster increased trade and cooperation among nations while not compromising or destabilizing any of the major regional or global security imperatives. New technology–security linkages have emerged because of the evolving nature of new and enabling technologies and new threats of asymmetric war fed by the alarming spread of terrorism and religious fundamentalism. This introductory chapter briefly reviews the technology aspects of this changing world scenario through an analysis of how technological advances, technology diffusion and technology controls have played a major role in the international security and stability calculus.

Chapter 2 outlines changes in security perceptions and highlights their interconnections with technology. The 21st century started out with a single superpower and an emerging world order, the contours of which are still unclear. Technology will play an important role in future regional conflicts. At the same time, the unavoidable spread of sensitive technologies has led to new concerns that they will reach irresponsible or unstable hands—giving rise to grave threats of asymmetric warfare. The relevance of conventional arms transfers for regional security is manifold and the role of technology in these fast-changing security perceptions and future conflicts is discussed. The proliferation of WMD technologies and concomitant concerns about the misuse of lethal or disruptive technologies have altered the security perceptions in this militarily unipolar world, which is also simultaneously emerging as an economically multipolar world. The non-proliferation focus has been more on control and less on removing the motivations for proliferation. The chapter therefore presents an examination of the causes of WMD proliferation and includes suggestions for revising non-proliferation priorities in consonance with the changing threat perceptions in order that the world can be better prepared for future challenges.

Technology control regimes have formed a vital component of national security strategies for controlling proliferation, particularly nuclear weapon and missile proliferation. Arms control initiatives have had many successes, but this does not mean that 'more of the same' will be the best solution for a long-term focus on progressive disarmament of all WMD. Furthermore, the negative impacts of export and technology controls on overall international cooperation, stability and economic progress, particularly from the point of view of the developing nations, have not received adequate appreciation.

Chapter 3 reviews the arms control, export control and disarmament issues from this different point of view in the context of the changing environment and presents a retrospective on the NPT and the 1996 Comprehensive Nuclear Test-Ban Treaty (CTBT)[26] as perceived by a developing nation such as India. The MTCR has emerged as one of the most discriminatory regimes with, understandably, mixed results. The impact of missile defence technology on the MTCR is therefore discussed in order to draw attention to the potential introduction to outer space of weapons that could trigger a new technology race for space defence. Finally, an assessment of the existing multilateral export control regimes is presented, bringing in the demand-side perspective to support the argument that future technology control systems must be more universal, transparent in nature and fair in approach.

Chapter 4 focuses on the subject of technology diffusion and discusses how the effects of globalization and market economic forces will pose new challenges for technology control regimes as further technology diffusion becomes inevitable. Advanced warheads, smart weapons, sophisticated cruise missiles and the use of outer space for surveillance and weapon control, as well as the revolutionary capabilities of network-centric warfare (NCW), have already altered the

---

[26] The CTBT was opened for signature on 24 Sep. 1996; it was not in force as of 1 Jan. 2004. The treaty will enter into force 180 days after it has been ratified by the 44 members of the Conference on Disarmament with nuclear power or research reactors on their territories, as listed in Annex 2 of the treaty. The text of the CTBT is reproduced in *SIPRI Yearbook 1997: Armaments, Disarmament and International Security* (Oxford University Press: Oxford, 1997), pp. 414–31.

basic tenets of defence and security. Technology–security linkages will become increasingly strong, and the international community will have to evolve new methods to deal with new patterns of technology proliferation to address the real concerns of uncontrolled spread and misuse. The security implications of some of the new technologies are discussed to highlight the increasing role of technology in the formulation of security strategy and military doctrines.

The overall aim of the report is to present a technology-oriented analysis of security, non-proliferation and arms control issues in the context of fast-changing international dynamics. It thus includes an appreciation of some of the emerging technologies and enabling high-value technologies that will undoubtedly be pursued by all sovereign nations, whether developed or developing. Technology gaps between progressive nations may indeed become narrower in some areas. This report attempts to articulate some contentious issues related to technology control and technology cooperation that deserve serious review if policies in this area are to remain relevant and effective in these changing times. In this light, the concluding chapter begins by presenting a summary of the main observations of this report and reflections on the future of technology controls in the context of the security challenges of the 21st century.

Global concerns and global dangers require global solutions through global cooperation. The future 'technology control' system must be based on a broader cooperative approach, with shared security perceptions. Chapter 5 therefore suggests a new approach to technology management that could facilitate a broader consensus on the control of and access to sensitive technologies through a universally acceptable, transparent and just methodology. An open system is envisaged as a valuable tool for assessing the non-proliferation performance of all participating countries, and thus contributing significantly to existing efforts to prevent technology misuse by terrorist, extremist or other 'rogue' elements. It would also foster better coordination and cooperation among all progressive nations, which will be vital to the optimal use of technology for the benefit of all. The management of future conflicts and control of potential dual-use technologies are likely to be two major areas of focus for future defence strategists and planners. They will certainly require pragmatic and innovative approaches and it is hoped that these will be along the lines suggested in this report.

2. Changing threat perceptions and proliferation concerns

I. The changing world order

In the absence of a counter-balancing force after the sudden end of the cold war, the security scenario around the world has undergone a profound change. Threat perceptions and national security interests now vary for different regions, depending on parameters that differ over time for different nations. For over 40 years the Soviet–US confrontation dictated global security perceptions. However, apart from a small number of war-alert situations such as the Cuban missile crisis of 1962, the two superpowers, even with their hair-trigger readiness for war, managed to avoid conflict through the well-established MAD doctrine—which also had a controlling effect on their regions of influence. It is interesting to note that since World War II over 200 smaller armed conflicts have been fought and almost all of these took place in the developing, or third world, regions. Most have been border disputes or intra-state conflicts and remained confined to their specific areas without escalating or endangering international peace and stability.[27]

Perceptions of threats to the national security of a sovereign nation are made up of a complex set of parameters. While border security is crucial for all sovereign nations, perceived threats from neighbours can vary enormously depending on the balance of power between nations, which is largely based on techno-economic balances. The weapon system capabilities and the techno-military strengths of adversaries are the prime factors that influence threat perceptions. However, external factors, such as the political and security environment in the region and relations between neighbours, as well as internal parameters, such as economic stability, the quality of governance

---

[27] Although there is no general agreement on the number of conflicts that have taken place since 1945, there is a consensus on the pattern of rise and decline in the number over time and the dominance of intra-state conflicts. See Seybolt, T. B., 'Measuring violence: an introduction to conflict data sets', *SIPRI Yearbook 2002: Armaments, Disarmament and International Security* (Oxford University Press: Oxford, 2002), pp. 81–96; and for conflict data from 1946 to 2002 see Uppsala Conflict Data Project, Department of Peace and Conflict Research, Uppsala University, Sweden, in association with International Peace Research Institute, Oslo, 'Armed conflict 1946–2002' at URL <http://www.prio.no/cwp/ArmedConflict/>.

and self-reliance in critical technologies, are also important factors influencing security perceptions.

Figure 2.1 gives a graphic representation of the various parameters that affect perceptions and definitions of 'national security'. The relative importance of these parameters differs for different nations. For example, the least developed nations may feel overwhelmed by problems of food, water and survival, while progressive and fast-developing nations strive for rapid techno-economic growth and a sense of self-reliance. A superpower such as the USA faces hardly any challenge to its basic security and well-being, but perceives the proliferation of military capabilities and disorder elsewhere as a possible future threat to its sense of stability and supremacy. Asymmetric threats from rogue states or terrorist groups therefore become the most imminent challenges to its security.[28] Unpredictable threats, ranging from the possibility of an accidental ballistic missile launch to a terrorist attack with CBW, also carry a much higher risk factor for a country such as the USA, which has a higher target value for terrorists compared to poorer developing countries. However, the level of inherent danger from such sudden, unpredictable threats is the same for all nations.

The scale and the sudden nature of the terrorist attacks on the USA of 11 September 2001 alerted the world to the common dangers of transnational terrorism and religious fundamentalism, which have been growing steadily over the past decade but not attracted the attention they deserve. It is arguable that the genesis of this violence can be traced back, in significant part, to the superpower tactics of using mercenary forces during the cold war years and the abrupt end of the cold war—providing a readily available, young mercenary cadre whose only self-definition lies in its capacity for violence. Of course, the root causes of fundamentalism can be attributed to the socio-economic gap between societies and disgruntlement linked to feelings of being exploited. Addressing root causes is important, but this will require a major reorientation in international thinking. In the immediate term, there is an urgent need to work for a concerted and cooperative approach to prevent terrorism from spreading further. This will require a focus not only on fighting terrorists but also on the effective elimination of the support infrastructures of all non-state players, as

---

[28] 'The National Security Strategy of the United States of America', The White House, Washington, DC, Sep. 2002, URL <http://www.whitehouse.gov/nsc/nss.pdf>.

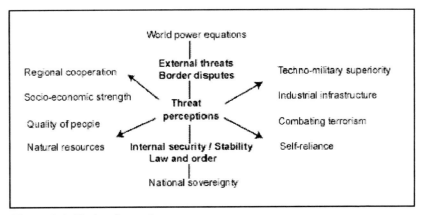

Figure 2.1. National security parameters

well as their state sponsors, for a period long enough for terrorist movements to lose momentum.[29]

Management of security perceptions linked to terrorist threats will face several new challenges because of the intrinsic unpredictability of the phenomenon. The most serious concern, of course, is that a terrorist organization might gain access to WMD. To address this concern comprehensively, all the possible routes of WMD proliferation to 'non-state actors' must be blocked, including the possibility of a desperate state with WMD technology assisting fundamentalist groups for a price, either economic or political. Theft of WMD or related technologies is also a possibility, either with or without the possible connivance of a rogue state. Hence, a terrorist group with access to 'garage-made' crude WMD is one of the most likely and dangerous scenarios. As is well known, a basic nuclear device is fairly easy to assemble with the right materials, and for intelligent and highly motivated groups, making rudimentary chemical or biological agents may be even more easily within reach.[30] The immediate need is therefore to enhance intelligence on the clandestine activities of terrorist

---

[29] Following the attacks on the USA the UN Security Council reaffirmed its unequivocal condemnation of terrorism and expressed its determination to prevent all such acts. UN Security Council Resolution 1373, 28 Sep. 2001. It also established the Counter-Terrorism Committee, made up of all 15 members of the UN Security Council, to monitor the implementation of Resolution 1373 and increase the capability of states to fight terrorism. See URL <http://www.un.org/Docs/sc/committees/1373/>.

[30] For more detail see Albright, D. and Higgins, H., 'A bomb for the Ummah', *Bulletin of the Atomic Scientists*, vol. 59, no. 2 (Mar./Apr. 2003), pp. 49–54.

groups and initiate preventive measures to thwart the intentions of such groups before they can strike. For long-term containment of terrorism, it will be necessary to isolate these groups financially and mount international counter-terrorism initiatives. Merely countering a region-specific threat will only cause terrorists to shift bases and change their modus operandi. Removing the threat of terrorism from international security calculations will be a major contribution to international peace and stability.

The increased availability of conventional weapons is equally disturbing because technology is imparting higher accuracy and lethality to these weapons and, as well as clandestine trade, economic compulsions are pushing them on to the open market. It should be recognized that the trade in small arms and light weapons (SALW) is largely responsible for the easy availability of such arms to terrorist organizations. If uniform national laws could be enacted for the mandatory incorporation of markings into all military products to preserve information such as type, number and details of manufacture this would significantly help to enhance transparency in such sales and transfers.[31] The use of markings would not only help in the subsequent tracking and monitoring of clandestine deals but also serve as a deterrent to states indulging in malpractice and provide incentives to be responsible in arms sales. New technology control methods, such as those suggested in chapter 5, could extract additional benefits from such transparency and thereby enhance international stability.

In the USA, the issue of homeland security appears, at present, to have overtaken strategic long-term security planning. However, the usual tendency to 'close the castle gates and raise the castle walls'[32] cannot be the solution in this age of globalization and interdependence. Nor will selective targeting of terrorists, even under a declared war on terrorism, be adequate—as the USA is painfully learning. Ter-

---

[31] Based on the Programme of Action that was agreed at the UN Conference on the Illicit Trade in Small Arms and Light Weapons in New York in July 2001, governmental experts have studied the issue of tracing illicit SALW. Subsequent to the report by the experts, the General Assembly in Dec. 2003 established an open-ended Working Group to negotiate an international instrument to enable states to identify and trace, in a timely and reliable manner, illicit SALW. The background to and activities of the open-ended Working Group on Tracing Illicit Small Arms and Light Weapons are available at URL <http://disarmament2.un.org/cab/salw-oewg.html>.

[32] Rogers, P., *International Security in the Early Twenty-first Century*, ISIS Briefing Paper no. 22 (International Security Information Service: Brussels, Feb. 2000), URL <http://www.isis-europe.org/ftp/download/bp-22.pdf>.

rorism and armed conflicts need to be dealt with differently.[33] In a world where the gap between the world's richest and poorest is increasing, violence and terrorism are unlikely to go away in the near future. Long-term solutions to transnational problems such as terrorism must be based on an intrinsically just and widely acceptable approach. For this to happen, the international community will have to take the lead in forging new alliances of nations for wider cooperative action against all forms of terrorism anywhere in the world. Universal non-acceptance of practices such as genocide and slavery has provided a basis to work for their elimination. Terrorism is another such crime against humanity that needs to be tackled with forceful, unanimous and global resolve.

A long-term perspective on security issues in the 21st century, however, reveals that conflicts arising from increasing economic polarization as well as the constraints of environmental limitations and the damage done by climate change will be major factors. Superimposition of these trends onto the increased availability of high-tech weapons could embolden a weaker group of states to take up arms against the powerful strong because of a collective 'anti-elite' feeling.[34] This could be further exacerbated because of the present tendency to maintain control through force or coercion, which addresses the symptoms but not the causes of instability. Today, the international divide is not only linked to political alignments or economic disparities but also to security-related issues such as the denial of technologies, imbalances in arms export policies, biases in trade practices and even environmental problems. Modern technology has allowed efficient dissemination of information around the world, creating a new ability among the poor majority to understand how badly they stand deprived of the things that are commonplace for the rich minority. It is therefore important to appreciate the need to change the old ways and initiate processes that can reverse socio-economic polarization and encourage environmentally sustainable development. This will be vital for international cooperation on future issues of global security and stability, and can be achieved without comprom-

---

[33] Stepanova, E., *Anti-terrorism and Peace-building During and After Conflict*, SIPRI Policy Paper no. 2 (SIPRI: Stockholm, June 2003), available at URL <http://editors.sipri.se/recpubs.html>.

[34] Rogers (note 32).

ising the focus on global non-proliferation and regional risk-reduction strategies.[35]

The global security environment today is much influenced by the perceptions of the most influential nation—the USA—and quite logically so. The most often articulated threat perception is therefore the possible misuse of WMD and the missiles used to carry them to their targets. The George W. Bush Administration's National Strategy to Combat Weapons of Mass Destruction[36] clearly enunciates the dangers but has a fairly narrow approach to addressing the real issues.[37] Although the term WMD encompasses nuclear, chemical and biological weapons, it should be recognized that CBW are legally prohibited under non-discriminatory treaties while possession of nuclear weapons is legal for the five permanent members of the UN Security Council (P5) and is also available, without violating any international treaty, to three other nuclear weapon-capable states outside the NPT.[38] A small number of other nuclear weapon aspirants continue to exist, and such aspirations will remain alive as long as nuclear weapons remain legitimate for some. There is a general agreement among the majority of nations that only the complete elimination of nuclear weapons can guarantee a lasting elimination of their threat. The P5, in recognition of this, have reaffirmed their commitment to the total elimination of nuclear arsenals in the foreseeable future.[39] However, many other non-proliferation initiatives must succeed first before the issue of universal nuclear disarmament can be given any realistic attention or a timetable.

Technology has made possible reductions in the size and weight of weapons and increases in their efficiency, range and accuracy. Future nuclear weapons may not even have to depend on ballistic missiles, although at present these may still pose the most threatening delivery

---

[35] Alfarargi, S., 'A view from the League of Arab States', and Adejumobi, S., 'A view from Africa', eds A. J. K. Bailes and I. Frommelt, SIPRI, *Business and Security: Public–Private Sector Relationships in a New Security Environment* (Oxford University Press: Oxford, 2004), pp. 235–53.

[36] 'National Strategy to Combat Weapons of Mass Destruction', The White House, Washington, DC, Sep. 2002, URL <http://www.whitehouse.gov/news/releases/2002/12/WMD Strategy.pdf>.

[37] Perkovich, G., 'Bush's nuclear revolution: a regime change in non-proliferation', *Foreign Affairs*, vol. 82, no. 2 (Mar./Apr. 2003), pp. 2–8.

[38] The 3 states are India, Israel and Pakistan.

[39] Johnson, R., 'The 2000 NPT Review Conference: a delicate, hard-won compromise', *Disarmament Diplomacy*, no. 46 (May 2000), URL <http://www.acronym.org.uk/dd/dd46/46 npt.htm>.

system. It is the realization of the limitation of the current arms control methods that has caused a change to US perceptions. Taking as its point of departure the fact that WMD could enable adversaries to inflict massive harm on the USA, the US National Strategy to Combat Weapons of Mass Destruction makes no mention of weapons possessed by US friends and allies, but it states that the USA 'will not permit the world's most dangerous regimes and terrorists to threaten us with the world's most destructive weapons'.[40] This reflects a more pragmatic way of looking at technological capability together with intentions, technologies and increased emphasis on intelligence.

The 1993 modification of the MTCR to include all unmanned vehicles intended for delivery of any WMD was itself a clear recognition of this enhanced threat perception. However, enforcing such wide-spectrum control mechanisms is fraught with enormous problems related to compliance verification, transaction identification, export control contradictions, pre-emptive actions and, of course, the management of damage inflicted before detection. Given such complexities, bringing all WMD-related missile proliferation down to a near-zero level seems almost impossible. Hence, nuclear weapons will continue to be a significant factor in the security calculations of most of the powerful and progressive nations, NWS or non-nuclear weapon state (NNWS), which have too much to lose from any nuclear disaster—intended or unintended.

The military operation in Iraq in March–May 2003 has altered the security geometry in many parts of the world, giving rise to new insecurities in some smaller developing nations. Conflicts of interest with a distant superpower are emerging as a major security concern because of the fear of even remote nations being coerced into situations conflicting with their individual national interests. This one event has created a security vacuum in international affairs because of what has been perceived as a serious degradation of the institution of the United Nations. Notwithstanding the questionable record of the UN in enforcing peace, it is the only manifestation of a fair and impartial world body designed to work on the universal merits of a given case rather than partisan considerations. During the first decade after the end of the cold war, there was much discussion about the possible emergence of an international system based on concepts of 'right and wrong'. However, a subjective and judgmental process

---

[40] 'National Strategy to Combat Weapons of Mass Destruction' (note 36), p. 1.

exposing weaker players to being identified as 'good or evil' by a powerful group of nations is now emerging. While the issues of unresolved regional conflicts will dictate region-specific threat perceptions, the security concerns of smaller or less powerful nations will invariably include concerns regarding outside intervention or coercion.

In contrast to the cold war years, the security balance of the 21st century has started to revolve more around economic competition and concerns over constraints on resources and the environment. The USA, as the sole techno-military superpower, may not perceive any real military threat, but may well weigh its own long-term interests against the natural aspirations of all the less powerful nations. Its rating of other nations will be heavily influenced by their techno-economic capacity to cooperate with, compete with or challenge its ultimate supremacy. Even in the context of the present US focus on homeland security, this demands that relentless efforts in military R&D must continue in the USA to keep its technological superiority far ahead of others. Other nations are unlikely to miss this significant message and future decades will witness a keen defence–technology race between the most technologically progressive nations, albeit at different levels of technological maturity.

By all indications, the 21st century could well emerge as an 'Asian century'. Asia is undergoing a dramatic transformation with high rates of economic growth and a major techno-economic revolution. Five of the eight nuclear weapon-capable states are in Asia. Asia supports more than 60 per cent of the world's population. China is slowly emerging as a world power, Japan is an important technology player that is aspiring to be a world-class techno-military force, India is already the world's fourth largest economy,[41] Russia appears to be on the path to recovery, and several of the smaller Pacific Rim nations have the capacity to make a significant contribution to the generation of wealth. The Asian region is thus pushing at the limits of possible growth and development, with all the associated sensitivities such as the demand for energy, the search for a technological edge, and the economic competitiveness that this implies. The security environment is therefore even more in a state of flux in this area than it is in the developed group of nations further west.

---

[41] India was the 4th largest economy in the world in purchasing power parity (PPP) terms, according to World Bank figures for 1999.

What happens in China and India, and between them, will have a profound effect in Asia—in economic, military and technological terms—and hence on world affairs. The region is also a hotbed of tension, with parts of Afghanistan and Pakistan still providing launch-points for transnational terrorism, often supported by narcotics trafficking and other illegal activities. Several low-intensity proxy wars are continuing to hold back the momentum towards progress, stability and peace in the region. However, the pressure for economic progress is strong. The development of the Association of South-East Asian Nations (ASEAN) and of the larger ASEAN Regional Forum (ARF) based on it[42] reflects a choice by many of the region's richest states to explore the potential of a European Union-style multilateral integrated approach for strengthening their existential security as well as their prosperity. Other sub-regions might profitably follow the same line.

The security scenario around the globe is thus undergoing a major transformation, with a number of push–pull factors, all interdependent and each with its own complexity. In addition, global institutions such as the UN and the World Bank are seen to be less effective than in the past and there appears to be a certain vacuum in the area of impartial and effective international coordination. This only adds to the insecurity of individual nations, particularly the progressive developing nations that are set on an ambitious economic development course and stand to lose a lot if they are hindered in the process. The paradigm shift in the world order towards a new, identifiable equilibrium is not yet complete. The USA is showing a preference for unilateralism. As a result, many multilateral initiatives for arms control, non-proliferation, environmental protection and so on may be compromised to some extent. However, in this interdependent world, unilateralism will eventually be forced to yield to cooperative multilateral approaches in order to tackle global issues. A clear appreciation of this inevitability and a fresh approach to deeper cooperation

---

[42] ASEAN was established in 1967 to promote economic, social and cultural development as well as regional peace and security in South-East Asia. The seat of the Secretariat is in Jakarta, Indonesia. The ARF was established in 1994 to address security issues. The ASEAN member states are Brunei Darussalam, Cambodia, Indonesia, Laos, Malaysia, Myanmar (Burma), the Philippines, Singapore, Thailand and Viet Nam. The member states of the ARF are the 10 ASEAN member states plus Australia, Canada, China, the EU, India, Japan, Korea (North), Korea (South), Mongolia, New Zealand, Papua New Guinea, Russia and the USA. (The ASEAN Post Ministerial Conference (ASEAN–PMC) was established in 1979 as a forum for discussions of political and security issues with dialogue partners. The member states of ASEAN–PMC are the 10 ASEAN member states plus Australia, Canada, China, the EU, India, Japan, Korea (South), New Zealand, Russia and the USA.)

among the larger group of responsible nations, quite different from the cold war approach of 'us versus them', should form the new world order for the 21st century.

II. The regional impact of conventional arms transfers

The main arms-exporting countries all maintain legal controls prohibiting the unauthorized export of weapons as well as associated items and technologies. Moreover, these countries all apply publicly accessible policy guidelines, which they use as the basis for their assessment of particular transfers. However, the export of conventional arms is a multi-billion dollar business activity that is governed for the most part by market forces and national political preferences, rather than by international arms control commitments.

A small group of suppliers dominates the international market for major, advanced weapons. In the five-year period 1998–2002 the five largest supplier countries accounted for about 80 per cent of all transfers of major conventional arms.[43] The USA was the largest supplier during that period with 41 per cent of the total share while Russia accounted for another 27 per cent. Since the initiation by the USA of the war on terrorism, there has been an expectation of increased production and trade in conventional arms to equip additional forces at home and abroad to combat terrorists. However, there is no satisfactory mechanism to monitor and distinguish between arms transferred to fight terrorism and arms transferred for use in regional conflicts.[44] If military anti-terrorist activities expand in the future and become long drawn-out operations there could be an increase in the volume of arms transfers or a change in the nature of the items that are transferred. Either development or both in combination might lead to destabilizing effects in certain regions of the world.

The imbalance between excess defence industrial capacity in most of the advanced industrial supplier nations and declining defence budgets linked to changing threat perceptions has to a certain extent been corrected by reductions in capacity and efforts to increase defence spending. However, many developing nations, with heightened security problems and a sense of lagging behind, are still hard-

---

[43] Hagelin, B. *et al.*, 'International arms transfers', *SIPRI Yearbook 2003* (note 6), p. 439.
[44] Hagelin (note 43).

Table 2.1. Five recipients of US aid related to the war on terrorism
All figures are in millions of euros.

Recipient	Before 11 Sep. 2001^a	After 11 Sep. 2001^b
Indonesia	49.9	76.9
Kyrgyzstan	35.3	87.8
Pakistan	3.5	1 293.5
Philippines	7.4	82.9
Uzbekistan	28.1	171.7

^{a} These figures include aid appropriated under the 2001 Foreign Appropriations Act.

^{b} These figures include aid in 2 supplementary appropriations acts passed after 11 Sep. 2001 and aid requested as part of the 2003 Foreign Appropriations Act.

Source: Gabelnick, T. and Schroeder, M., 'Guns R Us', *Bulletin of the Atomic Scientists*, vol. 20 (Jan./Feb. 2003), pp. 38–39.

pressed to continue with defence acquisitions to modernize and enhance their capabilities. Ironically, these countries have emerged as attractive and important markets for supplier countries that often benefit from selling arms to both sides caught up in regional tensions. It may therefore be surmised that regional conflicts which remain within limits serve the arms trade in the supplier group of nations well. It might even be queried whether the priority for arms export controls has shifted to managing regional conflicts according to the political priorities of supplier nations. Defence industries in the technologically advanced countries will remain heavily dependent on exports and it is the recipient countries that will keep feeding these industries. Export controls will continue to be the major control mechanism for containing transfers within levels that suit the arms suppliers but do not destabilize global security. Arms transfers and military aid are thus dictated by the changing political and strategic priorities of major supplier nations, and these may not always match the priorities for enhancing international peace and stability.

In May 1996 the United Nations Disarmament Commission agreed on guidelines for international arms transfers.[45] Moreover, the UN Register of Conventional Arms (UNROCA) has brought about some

[45] 'Guidelines for international arms transfers in the context of General Assembly Resolution 46/36 H of 6 Dec. 1991', Disarmament Commission 1996, substantive session, New York, 22 Apr.–7 May 1996, available at URL <http://projects.sipri.se/expcon/acn10.htm>.

transparency in this area.[46] However, there is no overarching international agreement on controlling such arms transfers. Of the 73 sensitive destinations that could be identified in the lists and guidelines of the four founding participants of the WA (Germany, Japan, the UK and the USA), only 28 were found to be common to all.[47] This is a testimony to the breadth and the depth of the conflicts of interest in this field.[48] While it would be difficult to agree on a global approach to conventional arms control, the fact that the arms industries of important nations are heavily dependent on arms exports is undoubtedly one additional complicating factor.

With the balancing effects of the cold war gone, numerous regional conflicts have emerged as flash points that can be fuelled by the influx of modern conventional arms. Once a regional situation shows signs of escalating, it usually attracts international attention—often leading to some moderating measures and possibly a UN arms embargo. However, the changed international dynamics and the rise of non-state actors make it necessary to re-examine these issues in order to evaluate the long-term impact of conventional arms transfers on regional and international stability. Table 2.1 illustrates how the USA, the single largest supplier in the global arms market, suddenly changed its focus to fighting terrorism, resulting in major changes to its priorities.

Major conventional weapon systems and platforms such as tanks, aircraft and ships are a significant component of arms transfers. Advances in technology are continuously enhancing the accuracy, lethality and range of conventional weapons. At the other end of the spectrum there seems to have been no diminution in the use of improvised explosive devices and highly effective modern explosives and small arms for asymmetric warfare by non-state actors, as repeatedly demonstrated by terrorist attacks worldwide. Ingenious use of common low-end technology as well as the unprecedented effectiveness of the latest modern conventional weapons pose serious threats because the combination of the two can have disastrous consequences. As a simple example, the threat of shoulder-fired missiles to commercial

---

[46] For an analysis of UNROCA and the limits of its effectiveness in influencing actual security processes see Wezeman, S. T., *The Future of the United Register of Conventional Arms*, SIPRI Policy Paper no. 4 (SIPRI: Stockholm, Aug. 2003), available at URL <http://editors.sipri.se/recpubs.html>.

[47] Greene, O., *Developing an Effective Successor to COCOM* (Saferworld: London, 1995).

[48] See section V of chapter 3 in this volume.

aircraft is now well known,[49] and the genesis of this threat can be traced back to the supply of such weapons to the Mujahedin fighters in Afghanistan.[50]

With rapid advances in technology and the changing dynamics of complex modern international society, the arms control community has the unenviable responsibility for balancing the political priorities of powerful nations with urgent security concerns and potentially dangerous possibilities arising from conventional arms transfers.

While in the past the risks associated with ballistic missiles have received much attention, recent trends clearly indicate that cruise missiles and even UAVs will probably overtake them in the field of short- and medium-range, high-precision attack.[51] In fact, with the probability of intercontinental war receding, export control systems focused primarily on long-range ballistic missiles will seem increasingly inadequate. It is technological sophistication and hostile intent that are emerging as the main concerns. In this context, the control of conventional weapons will assume increased significance.

US Tomahawk cruise missiles exemplify the real missile strength in the world and the use of UAVs in combat may transform the rules of the future use of conventional weapon systems. While the technology for cruise missile guidance is fairly complex and currently out of the reach of many countries, there are an estimated 75 000 anti-ship cruise missiles deployed by more than 70 countries in the world.[52] The technology behind UAVs is much simpler and nearly impossible to control. Apart from where the most exacting performance is required, UAVs will emerge as the low-cost weapon solution to be used in many innovative ways in regional conflicts. Given the unique capability of low-flying cruise missiles to evade detection by enemy

[49] Man-portable air defence systems (MANPADS) are surface-to-air missiles small enough to be fired from the shoulder or from a small stand. The portability of these weapons, and their potential effectiveness against large and slow aircraft such as civilian airliners, creates a significant risk of terrorist acquisition and use. There have been a number of cases of actual or attempted use of such weapons against civilian aircraft. Moreover, for Russia, the use of these weapons by opposition fighters in Chechnya has led to the loss of significant numbers of helicopters and fixed-wing military aircraft. Anthony, I., *Reducing Threats at the Source: A European Perspective on Cooperative Threat Reduction*, SIPRI Research Report no. 19 (Oxford University Press: Oxford, 2004).

[50] Ohlson, T., 'The trade in major conventional weapons', *SIPRI Yearbook 1982: World Armaments and Disarmament* (Taylor & Francis: London, 1982), p. 181.

[51] Gormley, D. M., 'New developments in unmanned air vehicles and land-attack cruise missiles', *SIPRI Yearbook 2003* (note 6), pp. 409–32.

[52] Gormley (note 51), p. 409.

radar and the high precision of weapon delivery, they are an effective weapon system that will be much sought after. The UAV may emerge as the poor man's cruise missile, with limited effectiveness but lending itself to ingenious uses for sometimes surprising results. The uncontrolled spread of this technology would be attractive to terrorist organizations and will be a major challenge to arms export control systems in the future.

III. Future conflicts and threat perceptions

Given the fast pace of techno-economic change and the political adjustments in progress since the end of the cold war, the future, although more stable at the global level, appears more unpredictable at the regional level. Barring catastrophic events beyond human control, such as major earthquakes or floods or a space disaster, the world is most likely to change in small, incremental steps—often with some limited predictability. Currently, the USA, in its new incarnation as the single superpower, has assumed a major role in international policing—but that could easily change if future US national interests dictate otherwise. If that were to happen, with the UN meanwhile marginalized to a moral advisory role, there could be a serious international vacuum because of the absence of proper international institutions to protect international peace and stability.

In terms of overall international security, whether the global situation becomes more or less violent will depend largely on how well regional tensions are managed and how far harmonious international cooperation can be established to prevent further alienation of the developing nations. Lack of proper threat assessment or the ineffective management of early danger signs related to regional conflicts could easily lead to a slide into anarchy, with a possible domino effect leading to increased levels of violence. However, any move towards enhanced trans-regional stability and peace will require a clear understanding of the issues and sustained international efforts to contain regional tensions before they spill over to other regions. Effective management of global security issues with so many interdependent variables and unpredictable factors will require wider cooperation among the majority of progressive responsible nations, to address root causes and not just symptoms or consequences.

While most of the smaller wars since World War II had only a marginal effect on international security and stability, the spread of sophisticated technology to many regions might mean that this will not be the case in future. Since regional conflicts will remain the drivers of instability, it is worth taking a closer look at the major current conflict situations. The Israeli–Palestinian conflict in the Middle East, the India–Pakistan tensions in South Asia and the North Korean threat to its neighbours in East Asia are three cases where conflicts involve countries with nuclear weapons and ballistic missile capabilities.[53] Hence, conflicts in these regions have the potential to escalate to other regions and affect international stability. There are also many unresolved conflicts in Africa, but these are more likely to remain confined to their particular regions and are thus less likely to affect overall international security.

The Middle East

The Middle East has always been of strategic importance because of its oil reserves and the strong US commitment to the security and prosperity of Israel.[54] The region has long been an area of tensions and regional armed conflicts. It has also emerged as a focal point for WMD competition, with several states interested in developing capacities designed to change the security and threat equations not only in the region but also in adjacent regions (see below). Because of wide techno-economic disparities between countries in conflict in the region, terrorism has gradually emerged as one of the established modes of conducting conflicts.[55] Other internal factors that affect security and stability include: (*a*) problems with slow economic growth and unemployment; (*b*) a breakdown of societal values and erosion of governmental control; (*c*) demographic changes leading to urbanization; and, of course, (*d*) the dominant forces of Islamic extremism and nationalism. The situation is further complicated because of the petrodollar riches of some states which, when com-

---

[53] The WMD aspects of these 3 situations are addressed in more detail in section IV below.

[54] Israel is the largest recipient of US aid, receiving around $3 billion a year, mostly as military aid. 'Doubts over US aid to Israel', BBC News Online, 20 Mar. 2003, URL <http://news.bbc.co.uk/2/hi/business/2867619.stm>.

[55] Examples include the transnational roles of Hizbollah and Hamas as well as the local use of terrorism techniques in the Israeli–Palestinian conflict. Stepanova (note 33).

CHANGING THREAT PERCEPTIONS 37

bined with a lack of strategic maturity or any long-term security focus, might provide an unstable mix for misplaced ambitions and bloody conflicts—not only in the case of Iraq. The pervasive insecurity among the states has led to high levels of military spending, often disproportionate to realistic national needs.[56] However, this report addresses the proliferation dynamics only in order to focus on the technological connotations of these regional conflicts.

Israel is the only de facto nuclear weapon state in the region and its forces, combined with its conventional superiority and the backing of the USA, remain a major source of concern and insecurity for other nations in the region.[57] There is therefore an acute awareness of the military utility of WMD as a low-cost alternative to seeking parity in the complex and expensive field of conventional weapon technology. WMD and missile capabilities are perceived as valuable both as a deterrent and as a potential tool for intimidation. The prestige value of 'WMD status' is also a major influencing factor for the dictatorial and feudal rulers of this region. Iran and Iraq have long had nuclear aspirations, Libya's moves in the same direction have recently been exposed, and others, such as Algeria, Egypt and Saudi Arabia, have at times shown an interest in acquiring WMD capabilities and/or delivery systems. The threats from Iraq and Libya now stand nullified after Iraq's enforced regime change and the British–US deal with Libyan President Muammar Qadhafi. Egypt, being one of the largest recipients of US aid, has for some time been open to US influence. However, CBW capabilities are well established in many parts of the Middle East and this may pose a major challenge to controlling their proliferation. In addition, concerns remain very much alive that one or two regional powers, such as Iran and possibly Syria, might succeed in achieving a nuclear weapon capability unless international pressure is brought to bear.[58] Beyond the WMD factor, the Middle East region

---

[56] For more information on military expenditure in these countries see Sköns, E. et al., 'Military expenditure', SIPRI Yearbook 2003 (note 6), pp. 322–26; and Omitoogun, W., 'Military expenditure in the Middle East after the Iraq war', SIPRI Yearbook 2004: Armaments, Disarmament and International Security (Oxford University Press: Oxford, 2004), pp. 381–88.

[57] For more on Israeli nuclear forces see Kristensen, H. M. and Kile, S. N., 'World nuclear forces', SIPRI Yearbook 2003 (note 6), p. 627.

[58] For the case of Iran see Kile, S. N., 'Nuclear arms control, non-proliferation and ballistic missile defence', SIPRI Yearbook 2003 (note 6), pp. 596–98. For the case of Syria see Nuclear Threat Initiative (NTI) and Center for Nonproliferation Studies (CNS), Monterey Institute of International Studies (MIIS), 'Syria profile: nuclear imports', July 2003, URL <http://www.nti.org/e_research/profiles/Syria/Nuclear/2083.html>; and NTI and CNS, MIIS,

continues to be the largest arms market in the developing world—accounting for nearly 60 per cent of world arms transfers. In this context it is interesting to note that Russia and the USA, with high stakes in the peace and security of the region, are, along with France, the dominant suppliers of arms, equipment and military technology to the region.[59]

Asia

Since Asia is expected by many to be the new strategic focus of international politics, it is important to understand how the same factors that contribute to its rise may also have the potential to fuel conflicts. Disparities among states are likely to increase because of differential rates of economic growth. However, the region is a complex mix of countries with varied cultures, faiths and social backgrounds—each governed by the compulsive forces of coexistence and cooperation operating across many regional boundaries for basic survival and development. The balance of power in the region will continue to be dynamic and unstable unless a clearer relationship emerges and/or a more cooperative approach to shared security is found.

India and Pakistan

The India–Pakistan conflict is a classic case of inherent imbalances created through artificial partitioning on religious lines when India gained independence from British colonial rule in 1947.[60] These imbalances have been further fuelled, as explained below, by outside interventions designed to use the geostrategic location of Pakistan for short-term objectives. The economic and political weakness of Pakistan both invited and necessitated its alignment with the USA during the cold war years when Indian–Soviet relations were generally good. China also used Pakistan's antagonism towards India to serve its own agenda of containing India by militarily assisting Pakistan. China is believed to have played a major role in providing nuclear technology

---

'Syria profile: nuclear overview', July 2003, URL <http://www.nti.org/e_research/profiles/Syria/Nuclear/index.html>.

[59] Hagelin (note 43), pp. 442–43.

[60] See Widmalm, S., 'The Kashmir conflict', *SIPRI Yearbook 1999: Armaments, Disarmament and International Security* (Oxford University Press: Oxford, 1999), pp. 34–37.

to Pakistan to counter India's superiority in conventional weapons.[61] Since its victory in the 1962 war against India, China has not regarded India as a major threat and has chosen to contain India by proxy means, so that China itself can focus on rapid techno-economic development.

Over the past decade India has emerged as a robust economy, growing steadily as a vibrant democracy despite a number of internal and external problems. Pakistan continues to suffer political convulsions as well as severe economic problems, and is generally regarded as having some way to go to consolidate its democracy under civilian rule. However, the international community continues to regard the two countries as a pair in a 'zero-sum' strategic calculation. Given the trends in the economic and techno-military fields, the present author believes that it is only a matter of time before the world recognizes India to be comparable more to China than to its smaller neighbour. Meanwhile, Pakistan's continued support for cross-border terrorism, under the assumption that its nuclear deterrence will inhibit India from taking strong action, is a perennial cause of concern to India and thus retains the potential to cause a major conflict in the future.[62] Although levels of militancy in Kashmir appear to have subsided and the people of Kashmir are showing a desire for peace, stability and progress, Pakistan's leadership has so far been unable to back down from its claims on Kashmir.[63]

The unresolved border tensions with Pakistan, combined with the nuclear missile capabilities across the borders, will continue to be of major concern in India's security perceptions, and also for other nations in the region. Another potential problem could be related to the introduction of significant amounts of modern high-tech conventional equipment and technology into the region. As pointed out above, under the immediate and narrow focus of the war against al-Qaeda terrorists, US military sales and assistance to Pakistan increased dramatically in just one year. In the long run, the Indian view is that Pakistan is almost certain to use its enhanced military capabilities to achieve regional gains against India.

---

[61] See Paul, T. V., 'China–Pakistani nuclear/missile ties and balance of power politics', *Nonproliferation Review*, vol. 10, no. 2 (summer 2003), pp. 21–29.

[62] Ramana, M. V. and Mian, Z., 'The nuclear confrontation in South Asia', *SIPRI Yearbook 2003* (note 6), pp. 195–212.

[63] See Ramana and Mian (note 62). In early 2004, however, the leaders of India and Pakistan began a process of dialogue focused primarily on reducing tensions over Kashmir.

China, mindful of India's growing regional stature, could intensify its efforts to contain India by fuelling low-intensity conflicts around India's borders. This would add new dimensions to potential future conflicts in the region. In terms of nuclear and missile capabilities, the existing deterrents, as well as the focus on economic development on both sides, are likely to hold for some time and thus any direct war or nuclear stand-off between China and India appears unlikely in the near future. While China continues to expand its techno-military capabilities with a focus on countering future threats from the USA,[64] it will also continue to watch India closely.

If the Asian region undergoes its own security transformation along these lines, India will be prompted to further sharpen its techno-economic deterrence against China. If Pakistan also continues to compete on a larger scale, depending on outside assistance or religious-based support from other Islamic nations, the chances for conflict in the Indian subcontinent could increase significantly. The situation therefore remains volatile and calls for urgent attention to forge regional confidence-building measures (CBMs), which should include a focus on nuclear-capable missile disarmament and a nuclear weapon-free zone among all the actors that have influence in the region.

The Korean peninsula

On the Korean peninsula, tight military alliances maintained a balance of forces between North and South Korea for more than 40 years. The sudden end of that balance with the end of cold war alliances, and the new ambitions of President Kim Jong-il's regime, have created the potential for a major conflict involving the use of WMD. Lack of information and understanding of the situation on the ground could lead to a crisis. The situation is further complicated by the alarming economic decline of North Korea coupled with the country's increasing isolation because of its belligerent policy on WMD proliferation. China and Russia seem unwilling to extend any long-term assistance and it is possible that systemic atrophy may lead to the collapse of the dictatorial regime and acute economic hardship leading to large-scale

[64] Jacoby, L. E., 'Current and projected national security threats to the United States', Statement for the record, Senate Select Committee on Intelligence, 11 Feb. 2003, p. 9, URL <http://www.fas.org/irp/congress/2003_hr/021103jacoby.html>.

migration from North Korea. Under such threats the regime might be tempted to change the regional balance of power through a suicidal full-scale use of force.[65] Technologies for the monitoring and verification of compliance with respect to NBC weapons will play a major role in controlling these situations, either with a view to averting war or to decisively eliminating the WMD threat.[66]

North Korea's missile capabilities and its history of trading in missile technology also add to concerns,[67] as the cash-starved regime could fall prey to the temptation to sell potentially dangerous equipment, even to non-state actors. Similarly, if North Korea succeeds in acquiring nuclear weapons, it could resort to nuclear blackmail to bargain for disproportionate gains. US intervention and negotiations have met with mixed results and, again, cooperation among the nations that have high stakes in the region will be vital for achieving lasting peace and stability.[68] Should the situation not be resolved, the possibility that Japan and South Korea will decide to develop their own deterrence capabilities will continue to haunt the region. This, in turn, could spur other countries in the region to further their military capabilities, thereby negating any chance of stability in the region.

Of the three major regional conflict situations discussed here, it is the assessment of the present author that tensions between India and Pakistan are least likely to escalate or spill over to other regions. The Far East and the Middle East situations are more serious in nature because the tensions are more volatile and any conflict would be capable of spreading and engaging larger powers.

Security and threat environments are thus undergoing a major change as the global threat patterns of the proliferation of dangerous

---

[65] Chanda, N., 'A quietly growing nuclear danger in North Korea', YaleGlobal Online, 28 Jan. 2003, URL <http://yaleglobal.yale.edu/display.article?id=824>.

[66] For the nuclear field see Zarimpas, N. (ed.), SIPRI, *Transparency in Nuclear Warheads and Materials: The Political and Technical Dimensions* (Oxford University Press: Oxford, 2003).

[67] See, e.g., Bermudez, J. S., 'A history of ballistic missile development in the DPRK', Occasional Paper no. 2, Center for Nonproliferation Studies, Monterey Institute of International Studies, Monterey, Calif., 1999, URL <http://cns.miis.edu/pubs/opapers/op2/index.htm>; and Wezeman, S. T., 'Suppliers of ballistic missile technology', *SIPRI Yearbook 2004* (note 56), pp. 545–49.

[68] In 2003 in Beijing 2 rounds of talks were held involving North Korea and the USA and aimed at resolving the crisis. The 1st round took place, with Chinese participation, on 23–25 Apr. and the 2nd round, on 27–29 Aug., included China, Japan, South Korea and Russia. Despite hints of diplomatic flexibility from both North Korea and the USA, the talks made little headway. Kile, S. N., 'Nuclear arms control and non-proliferation', *SIPRI Yearbook 2004* (note 56), p. 613.

weapons and terrorism, and the new parameters of techno-economic competition in an increasingly globalized world, are addressed. Containing proliferation and facilitating technological progress will require an objective understanding of the causes of proliferation and perhaps a reprioritization of proliferation concerns. On both counts technology will play a major role, whether for treaty compliance verification, threat anticipation or the selective but decisive control of sensitive technologies to avoid their misuse. The international community must unite and address the global influence of technology with the new maturity gained from the testing experiences of the past century.

IV. Causes of WMD proliferation

If changing threats are to be addressed comprehensively, it will be necessary to analyse the causes of WMD proliferation both in the regional context and in terms of technological interplay. This section therefore presents a closer look at the causes—rather than the effects—of proliferation. Proliferation of WMD is incontestably the most serious threat to international peace and stability, but perceptions of what proliferation is and how to control it are somewhat different for those with and those without access to technology. At the same time, the success of future initiatives to contain the proliferation of WMD capabilities will depend on the maximum convergence of views to facilitate wider cooperation among nations to counter the common threat.

The Soviet Union and the USA became NWS during the 1940s and the UK in the early 1950s. France followed with its own nuclear test in 1960 and China in 1964. These five nations are recognized NWS by virtue of the fact that they had conducted explosions prior to the negotiation of the NPT in 1967 and 1968. The first five NWS are therefore not classified as proliferators, although the unprecedented expansion of the nuclear weapon industry in these countries and the undisputed superior status of the NWS in world affairs provided a major incentive for all future proliferation.

The USA initiated the Atoms for Peace programme in the early 1950s to encourage the use of nuclear technology for peaceful purposes and as an incentive for others not to develop nuclear weapons.[69] This was perhaps the correct approach at the time. However, with hindsight it is easy to see that, given the political and security dynamics of the following years, it was unrealistic to expect that nuclear weapons would be contained within the five NWS. It is also interesting to note that, even within the 'club', the late entrants China and France only acceded to the NPT in 1992—24 years after it was opened for signature. These two parties were also the only nations that conducted additional nuclear tests before the conclusion of the CTBT negotiations.[70] The motivational factors behind the P5's choice to develop and maintain nuclear weapons are well known and provide valuable insights into the early stages of nuclear proliferation. However, nuclear proliferation tends to be discussed only as a post-NPT phenomenon: this report therefore focuses on the post-NPT period.

India, Israel and Pakistan are today de facto NWS. They are neither party to the NPT nor recognized officially as NWS. Instead, they are identified as nuclear proliferators because the NPT does not have any provision for accepting new states as NWS. India and Israel began establishing a national nuclear infrastructure in the 1950s and both had substantial indigenous technology capabilities in the 1960s.[71] Israel's efforts were made mainly in response to security concerns and Israel was the first state after the P5 to develop a nuclear weapon capability and a nuclear weapon arsenal. However, Israel has chosen to maintain official ambiguity (it will neither confirm nor deny) about its possession of nuclear weapons to avoid triggering an open nuclear race in the Middle East.[72]

---

[69] Weiss, L., 'Atoms for Peace', *Bulletin of the Atomic Scientists*, vol. 59, no. 6 (Nov./Dec. 2003), pp. 34–44.

[70] Ferm, R., 'Nuclear explosions 1945–94', *SIPRI Yearbook 1995: Armaments, Disarmament and International Security* (Oxford University Press: Oxford, 1995), p. 721; Ferm, R., 'Nuclear explosions 1945–95', *SIPRI Yearbook 1996: Armaments, Disarmament and International Security* (Oxford University Press: Oxford, 1996), p. 657; and Arnett, E., 'The comprehensive nuclear test ban', *SIPRI Yearbook 1995*, pp. 697–718.

[71] Perkovich, G., *India's Nuclear Bomb: The Impact on Global Proliferation* (University of California Press: Berkeley, Calif., 1999); and Cohen, A., *Israel and the Bomb* (Columbia University Press: New York, 1998).

[72] On 2 Apr. 1963, in a meeting with US President John F. Kennedy, Israeli Minister of Defence Shimon Peres said: 'I can tell you most clearly that we will not introduce nuclear weapons to the region, and certainly we will not be the first'. This formulation has subsequently become the standard Israeli response to questions about nuclear weapon capability. Cohen (note 71). A copy of the original Hebrew notes from the meeting form part of the archive maintained by the National Security Archive in Washington, DC. See 'Miscellaneous Hebrew documents', URL <http://www.gwu.edu/~nsarchiv/israel/documents/hebrew/>.

Israel's nuclear weapon programme was initiated in the late 1950s, at a time of close French–Israeli military cooperation. France is alleged to have provided Israel with design and manufacturing information for nuclear weapons.[73] The June 1967 war quickened the pace of Israel's nuclear weapon programme, leading to two improvised nuclear devices that were placed on operational alert. Israel has not officially carried out any nuclear tests, although it may have had access to information from US tests on some advanced devices.[74] This allowed Israel to build and stockpile a nuclear weapon inventory, estimated to be about 200 in number,[75] from about 690 kg of weapon-grade plutonium. Israel is the sixth nation to 'go nuclear' and the only nation in the Middle East with nuclear weapons. Israel appears to be moving towards a triad configuration of its nuclear forces and is reportedly attempting to acquire a survivable second-strike capability. Its legitimate security needs have generally been recognized by the NWS non-proliferation lobby, and Israel has never really been identified by them as a 'country of proliferation concern'. Israel has successfully maintained its nuclear ambiguity to date and, until it was linked with India and Pakistan after their 1998 nuclear tests, Western powers have largely avoided discussion of Israeli capacity under the non-proliferation agenda—presumably to help Israel maintain its nuclear ambiguity in the interest of Middle East peace efforts.

India has been a strong champion of non-proliferation and universal disarmament and its interest in nuclear technology was initially very much energy-oriented because of its historically heavy dependence on the import of crude oil. However, the lessons of external engagement by China and the USA in the 1971 war with Pakistan made India wise to the concept of nuclear deterrence and led directly to the decision to carry out the 1974 test.[76] Nonetheless, after demonstrating its nuclear

---

[73] In an article published in 1986 Francis Perrin, High Commissioner of the French Atomic Energy Agency in 1951–70, was quoted as saying that French and Israeli scientists worked closely together between 1957 and 1959 to design a nuclear weapon: 'we considered we could give the secrets to Israel provided they kept it a secret themselves'. Milhollin, G., 'Heavy water cheaters', *Foreign Policy*, vol. 69 (winter 1987/88), pp. 101–102.

[74] An Israeli scientist working at the US Los Alamos National Laboratory may have brought home expertise. Farr, W. D., *The Third Temple's Holy Of Holies: Israel's Nuclear Weapons*, Counterproliferation Papers, Future Warfare Series no. 2 (USAF Counterproliferation Center, Air War College, Air University: Maxwell Air Force Base, Ala., Sep. 1999), URL <http://www.au.af.mil/au/awc/awcgate/cpc-pubs/farr.htm>.

[75] Kristensen and Kile (note 57), p. 627.

[76] India conducted its first nuclear explosion, described as a 'peaceful nuclear explosion', on 18 May 1974. Barnaby, F., 'Nuclear-weapon proliferation', *World Armaments and*

technological capability through the peaceful nuclear explosion, and even in the face of increasingly complex security dynamics in the region, India refrained from further testing and from weaponizing its nuclear technology for over 20 years, which is unparalleled.

Until 1998 India maintained nuclear ambivalence in order to draw some deterrence benefits without kick-starting a regional nuclear weapon race. However, by the early 1990s India risked facing nuclear weapons in the hands of two neighbouring adversarial states (China and Pakistan) and had a history of wars with both countries. With new evidence of a nuclear-missile nexus between China and Pakistan,[77] India thus had little choice but to address its enhanced threat perceptions by establishing a minimum credible nuclear deterrence through the nuclear tests carried out in May 1998. According to some estimates, India is believed to have produced 240–395 kg of weapon-grade plutonium and a smaller amount of enriched uranium for the 30–40 warheads in its inventory.[78] While the 1988 nuclear tests attracted worldwide condemnation, it must also be recognized that the end to the prolonged nuclear ambiguity brought about a new level of pragmatism in the region and opened up the possibility of seeking stability through mutual deterrence, similar to the Soviet–US situation, but on a much smaller scale.

India announced a draft nuclear doctrine in August 1999,[79] accompanied by a unilateral moratorium on further tests and a no-first-use policy. India's National Security Council has been operational since 1999 and the National Security Advisory Board (NSAB) is also fully functional. The draft nuclear doctrine being followed in India, which was redefined in January 2003, was formulated by the NSAB.[80] For a survivable retaliatory strike capability India has decided to move towards a triad configuration with land-, air- and sea-launch capabilities. However, India continues to support the ideals of nuclear

Disarmament: SIPRI Yearbook 1975 (Almqvist & Wiksell International: Stockholm, 1975), pp. 16–22.

[77] Jones, R. et al., *Tracking Nuclear Proliferation: A Guide in Maps and Charts* (Carnegie Endowment for International Peace: Washington, DC, 1998), p. 136.

[78] For more detail see Albright, D., 'India's and Pakistan's fissile material and nuclear weapons inventories, end of 1999', Background Paper, Institute for Science and International Security (ISIS), 11 Oct. 2000, URL <http://www.isis-online.org/publications/southasia/stocks1000.html>.

[79] 'Draft report of National Security Advisory Board on Indian nuclear doctrine', 17 Aug. 1999, URL <http://www.indianembassy.org/policy/CTBT/nuclear_doctrine_aug_17_1999.html>.

[80] 'Nuke button rests in the PM's hands', *Indian Express*, 4 Jan. 2003.

non-proliferation and universal nuclear disarmament and maintains a strict self-imposed system of export controls to ensure the protection of all nuclear weapon technologies.[81]

Pakistan, on the other hand, started its nuclear weapon programme specifically in search of a weapon capability to counter India's stronger conventional strength. The resolve to 'go nuclear' came after its defeat in the 1971 war with India, and its accelerated efforts included acquiring weapons, components and technology from wherever possible by whatever means. Pakistan's logic in regarding its nuclear weapon capability as a deterrent may be legitimate, but its declared policy of refusing to rule out the first use of nuclear weapons to neutralize any conventional superiority of its adversary has lowered the nuclear weapon threshold in the region. The other strong motivation for Pakistan can be found in its apparent desire to shelter behind its nuclear deterrent while supporting militant Islamic extremism in the region, that is, because of the reduced likelihood of military retaliation by India, Pakistan is able to support cross-border terrorism in Kashmir with impunity.[82] This is the first case in history in which a so-called strategic capability is being used in the narrow context of a low-intensity border conflict being fought with the help of militancy and terrorism.

Pakistan is an interesting case for proliferation analysis, in terms not only of typical proliferation but also of shifting standards in international treatment. It is believed that its initial nuclear weapon programme relied heavily on the clandestine acquisition of key technologies from Germany, the Netherlands and the USA for the Kahuta nuclear facility.[83] After the 1977 Glenn–Symington Amendment to the 1961 Foreign Assistance Act, US military and economic aid to Pakistan was stopped because of its unsafeguarded uranium enrichment facility.[84] However, in 1979, in the wake of the Soviet occupation of Afghanistan, these restrictions were waived because Pakistan's support was strategically important for the USA in

---

[81] Detailed export control regulations and procedures are notified by the Government of India under the 1992 Foreign Trade (Development and Regulation) Act, which was updated in the Apr. 2000 Regulations for Control Over Export of 'Dual-Use' Goods of Indian Origin.

[82] Indian concerns about Pakistani support for terrorism in and aimed at Indian territories have been one of the main issues in bilateral security disputes. Ramana and Mian (note 62). They are being addressed in the recent negotiations aimed at relaxation of tensions.

[83] Spector, L. S. and Smith, J. R., *Nuclear Ambitions: The Spread of Nuclear Weapons 1989–1990* (Westview Press: Boulder, Colo., 1990), chapters 4 and 5.

[84] Spector and Smith (note 83).

its rivalry with the Soviet Union.[85] Pakistan did this with some distinction, creating the Mujahedin cadre that eventually merged with the Taliban groups. Notwithstanding various assurances to the USA about not producing weapon-grade uranium, Pakistan's nuclear weapon programme progressed steadily, producing enough nuclear material for its first weapon device by about 1986, ostensibly with significant help from China. However, events in Afghanistan forced the USA to dither between sanctions and support, as witnessed by the Pressler Amendment in 1985[86] and the sanctions announced in 1990, which were later relaxed and then lifted.

The 1998 nuclear tests that allowed Pakistan to claim a reliable weapon design for a 10- to 15-kt yield once again caused invocation of the Glenn–Symington Amendment sanctions by the USA. (These sanctions were also imposed on India.) However, the events of 11 September 2001 dramatically changed US policy towards Pakistan and the readiness of Pakistani President General Pervez Musharraf to assist the USA in fighting the terrorists based in the Afghanistan–Pakistan area was largely responsible for the sanctions being waived quickly. Even the sanctions on Pakistan that resulted from the army coup in October 1999 were waived immediately.[87] Sanctions against India were also lifted at the same time. This is a lesson on how the realpolitik of a superpower nation can be much more powerful than treaties and international ideology on non-proliferation and disarmament.

Highly autocratic North Korea, ruled by a dictator, is perhaps a more extreme version of the case of Pakistan: an economically weak

[85] The US Congress adopted the Symington Amendment to the 1961 Foreign Assistance Act in 1976: this amendment prohibits most US economic and military assistance to any country delivering or receiving nuclear enrichment equipment, material, or technology not safeguarded by the IAEA. Congress adopted the Glenn Amendment to the Foreign Assistance Act in 1977: this amendment prohibits US assistance to any non-nuclear weapon state (as defined by the Non-Proliferation Treaty) that conducts a nuclear explosion. These amendments did not apply retroactively to India or Pakistan. See Council for a Livable World, Arms Trade Oversight Project, 'India–Pakistan: sanctions legislation fact sheet', 11 June 2001, available at URL <http://www.clw.org/atop/restrictions_timeline.html>.

[86] The US Congress adopted the Pressler Amendment to the 1961 Foreign Assistance Act in 1985: this amendment banned most economic and military assistance to Pakistan unless the President could certify on an annual basis that Pakistan did not possess a nuclear device and that US aid would reduce the risk of Pakistan possessing such a device. Although Pakistan disclosed in 1984 that it could enrich uranium for nuclear weapons, and revealed in 1987 that it could assemble a nuclear device, the USA continued to certify Pakistan's non-nuclear status until 1990. Pressler Amendment sanctions were imposed against Pakistan in 1990, following the withdrawal of the Soviet Union from Afghanistan. Council for a Livable World (note 85).

[87] Ramana and Mian (note 62), p. 202.

country which became desperate to acquire nuclear weapons in order to gain an asymmetric capability to counter perceived threats from larger powers. North Korea is on the verge of becoming the ninth nation to acquire a nuclear weapon capability. Essentially a poor and isolated country, it may also attempt to use its nuclear and missile technologies for export or barter gains. North Korea has become an enigma for non-proliferation efforts. It is believed to have made enough plutonium at the Yongbyon reactor for five or six nuclear weapons[88] and is suspected of having established uranium enrichment facilities with clandestine help from Pakistan in exchange for missile components.[89]

North Korea acceded to the NPT in April 1985 but concluded an IAEA safeguards agreement only in April 1992, after the USA announced the withdrawal of its nuclear weapons from South Korea as part of an overall tactical withdrawal. However, the subsequent IAEA inspections led to fresh acrimony about unsafeguarded nuclear activity, which escalated further in 1994 when North Korea decided to de-fuel its 5-megawatt (MW) reactor, prompting the USA to propose a worldwide economic embargo. Visits by US President Jimmy Carter helped to diffuse the situation, and in 1994 North Korea agreed to dismantle the elements of its nuclear weapon-related activities in exchange for a number of energy- and security-related incentives from the USA.[90] According to recent reports, North Korea not only has declared the reprocessing of 8000 spent nuclear fuel rods[91] but is also suspected of having a second secret underground facility for the production of weapon-grade plutonium. After rejecting IAEA safeguards and serving notice of its withdrawal from the NPT, North Korea is now a major proliferation threat. It has used the USA's preoccupation with Iraq to its best advantage and has defied the international community with veiled threats that some interpret to include the possibility of a first strike with nuclear weapons against its neighbours if it is cornered.

---

[88] Albright, D. and O'Neill, K. (eds), *Solving the North Korean Nuclear Puzzle* (ISIS Press: Washington, DC, 2000), pp. 111–26.
[89] Hoagland, J., 'Nuclear deceit', *Washington Post*, 11 Nov. 2002, p. B7.
[90] Goodby, J. E., Kile, S. and Müller, H., 'Nuclear arms control', *SIPRI Yearbook 1995* (note 70), pp. 653–56.
[91] Kile, S. N., 'Nuclear arms control and non-proliferation', *SIPRI Yearbook 2004* (note 56), p. 612.

North Korea's carefully calibrated strategy of raising the ante against economic sanctions or military coercion is a new trend, but it is indirectly reminiscent of Pakistan's strategy of encouraging cross-border terrorism under the umbrella of a nuclear weapon threat. Such blurring of the nuclear weapon threshold is a dangerous trend that some dictatorial regimes may follow in the hope of exploiting an asymmetric advantage. This would seriously undermine the international momentum for non-proliferation and encourage other rogue or desperate states to strive to acquire nuclear weapons or other WMD.

Iraq, since the change of regime, will not be able to engage in nuclear weapon or other WMD proliferation for the foreseeable future. Attention has turned to how the US-led coalition or other players may be able to hold Iran back before it crosses the threshold.[92] Iran is a classic example of how a country can have serious security threats and legitimately want to enhance its defensive capabilities but also have questionable credentials for acceptance as a responsible democratic nation that can be trusted with sensitive technology. Iran was among the first group of countries to join the NPT as a NNWS when it ratified the treaty on 2 February 1970. However, the Iran–Iraq War, in which Iran was compelled to accept a ceasefire in August 1988, sharpened Iranian security awareness, particularly after the use of chemical weapons (CW) by Iraq and in view of the support that Iraq had received from various Western arms suppliers. In its neighbourhood, Iran perceives Israel as having not only nuclear weapons but also the solid support of the USA on all its security issues. The Middle East region is full of high-tech weapons imported from China, Europe, Russia and the USA. Nuclear weapons would therefore make strategic sense for Iran, and it has been suspected of secretly pursuing a nuclear weapon programme that, by one 2001 estimate, could lead to a weapon capability 'in the not too distant future'.[93]

---

[92] For more on Iran see, e.g., Kile, S. N., 'Nuclear arms control and non-proliferation', *SIPRI Yearbook 2004* (note 56), pp. 604–12.

[93] Kemp, G. (ed.), *Iran's Nuclear Weapons Options: Issues and Analysis* (Nixon Center for Peace and Freedom: Washington, DC, Jan. 2001), available at URL <http://www.nixoncenter.org/publications/monographs/IransNuclearWeaponsOptions.pdf>.

Between 1985 and 1992 Iran received a substantial amount of nuclear technology for its civilian programme.[94] It also received a positive report from the IAEA inspections that took place in November 1993.[95] In recent years, since Iranian President Mohammad Khatami came to power in 1997, Iran has worked hard to improve its world image and it stands to lose a lot if it is caught pursuing nuclear weapons in violation of its NPT commitments. The reactors that Iran received from China and its current contract with Russia for a 1000-MW reactor are under IAEA safeguards.[96] However, suspicion about Iran's objectives drove recent international attempts to prevail on it to sign an IAEA Additional Safeguards Protocol (allowing tighter inspections), although different interpretations of the nuclear 'freeze' which Iran offered as part of the same package are still causing concern.[97] Learning from the earlier experiences of Argentina and Brazil, where nuclear weapon aspirations were given up in response to incentives, it is possible that concerns about the potential military applications of the Iranian nuclear programme could yet be allayed.

These cases demonstrate the various causes of and motivational factors behind the complex issue of nuclear proliferation. It is clear that residual nuclear weapon aspirations outside the recognized NWS are very much based on regional security factors, and perhaps also influenced by a perception that nuclear technology capabilities make a country more important in international relations.[98]

Although the above commentary focuses on known horizontal proliferation cases, the successes of the non-proliferation regime in limiting the number of NWS and persuading countries such as Argentina,

---

[94] Albright, D., Berkhout, F. and Walker, W., SIPRI, *Plutonium and Highly Enriched Uranium 1996: World Inventories, Capabilities and Policies* (Oxford University Press: Oxford, 1997), pp. 359–60.

[95] In Dec. 1993 the IAEA reported that Deputy Director General for Safeguards Bruno Pellaud and his team had 'found no evidence which was inconsistent with Iran's declaration that all its nuclear activities are peaceful'. Nuclear Threat Initiative and Center for Nonproliferation Studies, Monterey Institute of International Studies, 'Iran profile: nuclear chronology 1993', Nov. 2003, URL <http://www.nti.org/e_research/profiles/Iran/1825_1870.html>.

[96] Feldman, S., *Nuclear Weapons and Arms Control in the Middle East* (MIT Press: Cambridge, Mass., Jan. 1997).

[97] International Atomic Energy Agency (IAEA), 'Iran to sign Additional Protocol and suspend uranium enrichment and reprocessing', IAEA Press Release 2003/13, Vienna, 10 Nov. 2003, URL <http://www.iaea.org/NewsCenter/PressReleases/2003/prn200313.html>.

[98] For a detailed treatment of the subject see Cirincione, J. et al., *Deadly Arsenals: Tracking Weapons of Mass Destruction* (Carnegie Endowment for International Peace: Washington, DC, 2001).

Brazil and South Africa to give up their nuclear weapon aspirations are noteworthy and can provide valuable lessons on how best to control future proliferation. Clearly, security assurances combined with economic assistance and high-technology cooperation can provide attractive incentives and serve specific non-proliferation initiatives in the future. Germany, Japan and South Korea also provide good examples of cases in which regional history and international dynamics have helped nations reject any search for a nuclear weapon capability. However, existing nuclear weapon arsenals held by the known NWS remain the most influential motivation for others to want to acquire nuclear weapons.

India's overt nuclear weapon status should be recognized as one of the major failures of the nuclear non-proliferation movement—a failure not so much in restricting India as in failing to address the genuine security concerns in the region. While the present efforts being pursued for various confidence-building and risk reduction initiatives are important to reduce the potential hazards from maintaining existing nuclear weapon stockpiles, the ultimate long-term gain from a possible nuclear weapon-free world could have a staggering impact in this new era of fast-changing threat perceptions and technological sophistication. As discussed in chapter 3 of this report, universal nuclear disarmament, however impractical and idealistic it might seem today, must remain the ultimate goal in the interests of world peace and stability.

V. Prioritizing proliferation concerns

The WMD–missile combination offers enormous deterrence value. A nation that perceives a strong need for this unique advantage will aspire to acquire it as long as it is seen as legitimate for others to have it. However, nuclear technology for the development of effective weapons is complex and expensive, requiring elaborate infrastructure and testing. Hence, it cannot be developed in a short time, nor can all countries afford it. Therefore, nuclear weapon activity is usually detectable before a country reaches threshold levels. Some technology diffusion is unavoidable—increasingly so in an interdependent world. Any effort to contain the dangers of misuse of nuclear technology has to factor in this reality. Countries with disproportionate riches (e.g., from petrodollars) or that receive outside assistance based on ideological or political alignments, or even on religious faith, can acquire nuclear weapons from an outside source. Such proliferation is, in its initial stages, more difficult to detect.

The development of CBW requires relatively simple technologies. Particularly for biological weapon (BW) precursors, small-scale efforts are often adequate and can therefore be pursued in conditions of high secrecy. CBW are also more difficult to detect or monitor for verification purposes. The non-discriminatory international treaties, the BTWC and the CWC, are proof of wider international agreement on a total ban on these weapons. However, some key countries of concern remain outside these treaties and the arms control community has to face the challenge of bringing these hesitant countries into the larger universal consensus.[99]

Economic and political dimensions have played a major role in the exchange of nuclear fissile materials between nations—often undercutting long-term non-proliferation objectives for immediate short-term gains. International responses to proliferation are not always the same because not all cases are seen as equally dangerous for international stability. It is arguable that a responsible democratic nation's acquisition of a defensive nuclear weapon capability is perceived as less dangerous to international peace and may even strengthen regional stability in the long term. If unstable dictatorial states or non-state terrorist organizations were to acquire offensive nuclear weapon capabilities, the results would be far more dangerous and destabilizing.

The technology for space launch vehicles (SLVs) and hypersonic cruise vehicles (HCVs) will continue to progress and this will make conventional propulsion and guidance technology too commonplace to control. Given the increasingly easy access to technologies for weapon delivery systems, future threats, in addition to ballistic and cruise missiles that are listed in the MTCR Annex,[100] may include the use of short-range missiles, anti-ship or cruise missiles, UAVs or even aircraft on suicide missions.

WMD technology controls and export control regimes of the future will have to concentrate much more on warhead technology than on delivery systems. Legitimate types of warheads for use in declared

---

[99] Zanders *et al.* (note 21). Complete lists of BTWC parties, signatory and non-signatory states are available on the SIPRI CBW Project Internet site at URL <http://projects.sipri.se/cbw/docs/bw-btwc-mainpage.html>.

[100] See notes 13 and 15.

war situations must be reduced to a minimum, with universal de-legitimization of all WMD that pose dangers only to human life without any rational military application.

There now being little chance of any UN member state using CBW against another, it should be possible to achieve universal CBW disarmament through an aggressive implementation of the existing conventions towards destroying all such weapons and closing down all production facilities. International attention must concentrate particularly hard on the few reluctant countries that are believed to have developed or to be developing chemical or biological weapons but have yet to join the BTWC and the CWC. Minimal defensive R&D could be conducted by a multinational team under a UN-approved agency with due transparency to maintain technical competence for inspection and verification as well as confidence among parties to the conventions.

Given the continuing importance attached to nuclear weapons by powerful nations, universal nuclear disarmament may not be possible in the short term. Hence, the control of nuclear weapons and material becomes an important issue. The CTBT remains a potentially valuable disarmament tool, and fresh efforts are needed to maintain and extend international support for it. A fissile material cut-off treaty (FMCT) would be another valuable step towards universal, and therefore balanced, disarmament.[101] The 2005 NPT Review Conference should find ways to accept the reality of existing NWS and build in flexibility to consider the legitimate entry of additional responsible states, should regional security environments in the future justify such a response. This would make the NPT more pragmatic, less attached to history and more effective in the future.[102]

In the light of the changed world scenario and an awareness of the real dangers of sensitive technology being accessible to terrorist or fundamentalist groups, it should now be possible to revisit the whole issue of controlling the proliferation of WMD and sensitive dual-use

---

[101] A fissile material cut-off treaty is another future treaty under discussion at the Conference on Disarmament (CD) for control of fissile materials. While this could prove a very effective tool to check further nuclear proliferation, it could also be a major catalyst for progressive step-by-step disarmament. However, FMCT negotiations have not started because of disagreements between CD members, including China's linking of FMCT objectives to resolution of the issues of missile defence and weapons in space. For more detail on the FMCT see the Internet site of the Federation of American Scientists, 'Fissile material cut-off treaty [FMCT]' at URL <http://www.fas.org/nuke/control/fmct/index.html>.

[102] These issues are discussed in detail in chapter 3 in this volume.

technologies. The real dangers are from asymmetric warfare by terrorist or fundamentalist groups—with or without the support of state actors. If religious fundamentalism, terrorist tactics and mercenary practices are allowed to grow, not even the states that currently support such activities will be safe. Prohibiting proliferation of WMD and the related technologies that allow such groups the access to such potentially dangerous technologies obviously becomes the first global priority and challenge today, a cause in which the USA—as world leader—can expect universal support.[103] A nexus of terrorists and the underworld would be the most disastrous recipe for the future of international peace.

The focus of international non-proliferation efforts and the attention of the arms control community must therefore be changed, in line with the technological and strategic options of the future. A possible re-prioritization of the focus of non-proliferation is as follows.

1. The focus of nuclear non-proliferation must now be narrowed to the prime threats from rogue states and fundamentalist organizations. Technology trends indicate that a sharper focus on warhead technologies rather than the full range of delivery systems may produce better results. Simply widening the scope of related technology controls may dilute such a focus.

2. Verification technologies will assume increased importance and thus will require wider cooperation among groups of nations. Intrusive inspections or discriminatory controls can easily add to international tensions, mutual suspicions and non-cooperation. The future success of non-proliferation will depend on a regionally sensitive and balanced approach to technology management rather than the extension of old-style technology controls.

3. A new and dangerous emerging trend is nuclear sabre-rattling by weak dictatorial regimes to blackmail larger powers. This must be discouraged outright, lest it acquire de facto international acceptance and thereby lower the threshold for the use of nuclear weapons.

4. All sovereign states must unambiguously confirm the total banning of all CBW activities, including R&D, production and procure-

---

[103] The USA has asked the UN Security Council to adopt a resolution calling on all members of the UN to 'criminalize the proliferation of weapons . . . of mass destruction'. 'President Bush addresses United Nations General Assembly', The White House, Office of the Press Secretary, Washington, DC, 23 Sep. 2003, URL <http://www.whitehouse.gov/news/releases/2003/09/print/20030923-4.html>.

ment. All R&D for countering CBW threats should be transparent and shared under the control of a universal institution, in order that individual states are not tempted to undertake separate R&D. The successful implementation of the BTWC and the CWC will be an enormous boost to nuclear non-proliferation and even to steps towards near-total disarmament.

5. There should be deliberate, gradual and progressive movement towards universal disarmament of nuclear weapons through a process of reduced dependence on and progressive devaluation of these weapons by the leading industrial nations. A no-first-use declaration by all NWS could be a good starting objective, leading to declarations of no-use (except with international consensus in extreme cases of threats to global security). The way ahead is discussed further in chapter 5.

VI. Future challenges

The biggest challenge to WMD proliferation in the immediate future will be the successful implementation of the BTWC and the CWC, in a comprehensive manner that would enable universal disarmament in these two areas to be achieved within the next few decades. Such a success would be a huge boost to the global non-proliferation agenda and would also provide valuable experience for the future.[104]

---

[104] Great emphasis is currently being placed on trying to ensure that the prohibitions contained in the BTWC and the CWC are effectively extended to non-state actors (i.e., individuals and groups). To achieve this, the conventions require that the parties pass national laws, including the enactment of penal legislation. Most of the parties have, historically, either not possessed national implementing legislation or possessed legislation that was not sufficiently comprehensive. The current efforts to address this problem enjoy a high level of political support and engagement in large part because of concerns about possible terrorist threats. It is also important that the parties remain actively engaged by, e.g., regularly considering whether and how relevant scientific and technological developments are reflected in treaty implementation. The development and synthesis of toxic chemicals, e.g., have been revolutionized partly through advances in combinatorial and computational chemistry and microarray processing technologies. Another area of possible concern is whether and how activities undertaken as part of national bio-defence programmes, including those involving the development of non-lethal weapons and incapacitants, should be covered by the conventions. Finally, it is important that institutional memory and expertise be retained and that technical and political aspects of treaty implementation be, to the extent possible, kept separate. Hart, J. et al., SIPRI, 'Maintaining the effectiveness of the Chemical Weapons Convention', Paper presented at the Organisation for the Prohibition of Chemical Weapons, The Hague, Netherlands, 8 Oct. 2002, available at URL <http://projects.sipri.se/cbw/research/cwc_policy paper2.pdf>; Zanders, J. P., Hart, J. and Kuhlau, F., SIPRI, 'Biotechnology and the future of the Biological and Toxin Weapons Convention', Fact sheet distributed at the Fifth Conference of the States Parties to the Biological and Toxin Weapons Convention,

In the nuclear field, preventing Iran from abandoning the NPT to 'go nuclear' will be a real test of the international non-proliferation community. Failure to achieve this would set further precedents for nations to renege on international treaty commitments. International success, on the other hand, would by analogy strengthen the NPT and help to limit further proliferation.

The other challenge for the NPT regime will be to find ways to accept the realities of the 21st century. Continuation of old norms and definitions could make the NPT anachronistic or harm its credibility. The real challenge for the arms control community will be to create conditions under which all nuclear weapon-capable states should commit to universal nuclear disarmament in a step-by-step manner. This is a tough challenge, but it is an essential tenet of non-proliferation that cannot be abandoned.

As suggested above, working towards a comprehensive no-first-use commitment by all the NWS might be a first step. The next step, depending on how technology-management initiatives succeed in prohibiting the possible misuse of other potentially dangerous technologies, would be a dialogue on no-use or disarmament, in the interests of global stability. In the opinion of many, a future global energy solution will largely involve nuclear power, and thus international cooperation in this field is vital for ultimate global energy conservation and the avoidance of an ecological disaster. Universal nuclear disarmament could help create the right framework for nuclear energy cooperation based solely on technological and commercial parameters. In addition to existing measures to prohibit the further spread of nuclear weapons, universal respect for the CTBT and the conclusion of an FMCT could not only limit the spread of nuclear weapons to new aspirants, but also limit R&D of a new generation of weapons that might bring about new dangers. If the world resigns itself instead to further applications for nuclear weapons, notably in space, it could soon find itself back under the constant threat of 'nuclear winter' or other unknown dangers.

The use of chemical weapons in the 1980s added a new dimension to proliferation concerns that had earlier related mainly to nuclear

19 Nov.–7 Dec. 2001, Geneva, Switzerland, available at URL <http://projects.sipri.se/cbw/research/biotechnologyfactsheet.pdf>; Hart, J., Kuhlau, F. and Simon, J., 'Chemical and biological weapon developments and arms control', *SIPRI Yearbook 2003* (note 6), pp. 659–66; and Lewer, N. (ed.), *The Future of Non-Lethal Weapons: Technologies, Operations, Ethics and Law* (Frank Cass & Co.: London, 2002).

weapons and their delivery systems. Since the collapse of the Soviet Union and the 1991 Gulf War, there has been further evidence of substantial programmes related to CBW development and stockpiling of CBW warheads, leading to greater consciousness of non-nuclear proliferation dangers. Accordingly, proliferation concerns have been revised to include all types of warheads capable of delivering WMD (nuclear, biological, chemical and radiological) irrespective of any weight or volume qualification. The MTCR has also been modified to include UAVs known or suspected to be for WMD delivery without any range or type qualification.

The challenge for the future will be to deal with the new dimensions of the WMD threat. The major threat to security and stability is now from sub-national or non-state players which might acquire and use WMD, with or without covert support from irresponsible governments. The perceived dangers of unstable or weak nations acquiring WMD technologies have also become more serious because of the increased accessibility of these sensitive technologies and materials, the concern that they might proliferate to potentially hostile or unstable owners, and the risk that such states or groups might resort to nuclear blackmail.

The USA, as the sole superpower, is progressing quickly with technological advances while all others are lagging behind by margins so large that not even its closest allies can catch up in the foreseeable future. The West's non-proliferation focus and export control agenda are being given a new orientation in terms of 'bad guys' and 'good guys'—a judgement that can change for political reasons and over time. This risks diverging from and compromising institutional agendas designed to curb the long-term impact of proliferation on global security and stability.

The other global challenge is the serious religious divide that is emerging, with Islamic groups feeling targeted almost as a class in the war against terrorism. The international focus on immediate problems has detracted from the long-term vision of the need to build more harmony among the nations and peoples of the world. It is possible that the world is moving into a period of greater tension and more proliferation, requiring more controls and more international policing, which, in turn, risk leading to sharper divisions and still greater tensions. The USA has a major responsibility to become the role model for the future. It will have to learn the responsibilities of a 'big brother' among nations, behaving with a fairness and magnanimity that behoves a superpower. No country understands technology better than the USA. It would therefore be natural to expect the USA to evolve techniques for the effective management of future technologies in order to foster global benefits for all humanity, while protecting all from the mindless misuse of technology.

3. A different perspective on arms control and export control regimes

I. Arms control and disarmament

The origin of arms control can be attributed to the basic national security imperative to reduce both the incidence of armed conflict and the potential for damage in a conflict situation. Technological advances have been instrumental to all human development but, in the same way as the industrial revolution mechanized warfare, they have created modern weapons with greater potential to cause damage. To capture the process in its conceptual stage, it is possible to reflect back to the 1899 and 1907 Hague Conventions, which ban the use of 'dumdum' bullets and the use of 'poison or poisonous weapons', respectively,[105] and the 1925 Geneva Protocol for the Prohibition of the Use in War of Asphyxiating, Poisonous or Other Gases, and of Bacteriological Methods of Warfare.[106] These early efforts to limit the development and acquisition of dangerous weapons came to be identified as arms control.

The advent of nuclear weapons vastly enhanced the potential for mass destruction. During the 1950s several new initiatives were set in motion to prevent the spread of such devastating weapons, and arms control started to assume a high priority on the national security agenda, particularly for the two superpowers of the time. Over the years the scope of arms control has been extended to include all WMD, their delivery systems and the sensitive dual-use technologies that contribute to such capabilities.

The concepts, procedures and enforcement structures for the implementation of arms control policies continued to evolve during the cold war years to produce the present-day control systems. These are essentially technology control mechanisms to prevent or limit the proliferation of dangerous weapons. The international treaties and informal multilateral arrangements that have evolved in the past dec-

---

[105] For more on Declaration IV, 3 of the 1899 Hague Peace Conference and Convention IV of the 1907 (Second) Hague Conference see Goldblat, J., PRIO and SIPRI, *Arms Control: The New Guide to Negotiations and Agreements* (SAGE Publications: London, 2002), Part I (Analytical survey), p. 280.

[106] Goldblat (note 105), pp. 135–37.

Table 3.1. Membership of multilateral weapon and technology transfer control regimes, as of 1 January 2004

State	Zangger Committee[a] 1974	NSG[b] 1978	Australia Group[a] 1985	MTCR[c] 1987	Wassenaar Arrangement 1996
Argentina	x	x	x	x	x
Australia	x	x	x	x	x
Austria	x	x	x	x	x
Belarus		x			
Belgium	x	x	x	x	x
Brazil		x		x	
Bulgaria	x	x	x		x
Canada	x	x	x	x	x
China	x				
Cyprus		x	x		
Czech Republic	x	x	x	x	x
Denmark	x	x	x	x	x
Finland	x	x	x	x	x
France	x	x	x	x	x
Germany	x	x	x	x	x
Greece	x	x	x	x	x
Hungary	x	x	x	x	x
Iceland			x	x	
Ireland	x	x	x	x	x
Italy	x	x	x	x	x
Japan	x	x	x	x	x
Kazakhstan		x			
Korea, South	x	x	x	x	x
Latvia		x			
Luxembourg	x	x	x	x	x
Netherlands	x	x	x	x	x
New Zealand		x	x	x	x
Norway	x	x	x	x	x
Poland	x	x	x	x	x
Portugal	x	x	x	x	x
Romania	x	x	x		x
Russia	x	x		x	x
Slovakia	x	x	x		x
Slovenia	x	x			
South Africa	x	x		x	
Spain	x	x	x	x	x
Sweden	x	x	x	x	x
Switzerland	x	x	x	x	x

State	Zangger Committee[a] 1974	NSG[b] 1978	Australia Group[a] 1985	MTCR[c] 1987	Wassenaar Arrangement 1996
Turkey	x	x	x	x	x
UK	x	x	x	x	x
Ukraine	x	x		x	x
USA	x	x	x	x	x
Total	**35**	**40**	**33**	**33**	**33**

Note: The years in the column headings indicate when the export control regime was formally established, although the groups may have met on an informal basis before then.

[a] The European Commission participates in this regime.

[b] The Nuclear Suppliers Group. The European Commission is an observer in this regime.

[c] The Missile Technology Control Regime.

Source: Anthony, I. and Bauer, S., 'Transfer controls and destruction programmes', SIPRI Yearbook 2004: Armaments, Disarmament and International Security (Oxford University Press: Oxford, 2004), p. 738.

ades for controlling or regulating arms exports are among the main instruments of arms control. Organizations such as the International Atomic Energy Agency and the Organisation for the Prohibition of Chemical Weapons were established under the aegis of the UN to administer and implement the relevant treaties on nuclear and chemical weapon technologies, respectively. At the conceptual level there is a fair degree of clarity and logic to the control regimes, but this is often undermined in practice because of the inevitable linkages between export control decisions and the political priorities of nations.

The cold war period was also responsible for another form of technology control: embargoes on the transfer of dual-use technology and products, specifically oriented to deny technology and knowledge to an adversary. This was the essence of COCOM, which became irrelevant after the end of the cold war. The Wassenaar Arrangement was configured by the original COCOM members as a successor export regulation regime but with an expanded agenda—to strengthen national export controls for conventional arms transfers and dual-use technology items through the exchange of information between participating states.[107] Under the WA the decision to transfer or deny any controlled item remains the responsibility of the participating states. Unlike COCOM, there are no case-by-case prior reviews of proposed exports to proscribed destinations, and no country can veto a proposed export. Nevertheless, despite the changed ideological context, most demand-side nations still view the WA as the successor to COCOM because it has developed control lists that include items not related to WMD arms control and non-proliferation treaties. While the documents that established the WA make clear that no country is an explicit target of the arrangement, participating states do use the forum to influence one another and each of them has a national view on which states are a cause for concern.[108]

Over time, the distinction between arms control and export control has become blurred, although they are intrinsically quite different— even though they may be considered as complementary. Arms control takes place among willing parties, whereas export controls are exercised by participating supply countries and the targeted countries have no role or say in the matter.

The USA was generally technologically ahead of the Soviet Union during the cold war, and denial of technology to the Soviet Union served the distinct purpose of maintaining the West's technological edge. The USA's technology quest was spurred by the first Sputnik launch by the Soviet Union and considerable US efforts were pressed into service at the time to discover the secrets behind the Soviet Union's lead in space technology. Espionage for technology information has always been carried out by both sides and every nation has drawn lessons about the need to protect its technological superiority by every possible means. It is interesting to note that over the years the practice of technology denial has been accepted as a perfectly legitimate exercise. By inference, could it not then also be posited that seeking technological advantage for reasons of national security by all legitimate means should be considered an equally valid goal? A nation devoted to acquiring the military capabilities deemed necessary

---

[107] COCOM (note 22) and the Wassenaar Arrangement (note 24) are discussed in more detail in section V below.

[108] E.g., from the outset the USA has consistently argued that particular restraint should be used in assessing transfers to Iran, Iraq, North Korea and Libya. See 'Statement of John D. Holum, Senior Adviser for Arms Control and International Security, Department of State, before the Senate Governmental Affairs Committee Hearing on the Wassenaar Arrangement and the Future of Multilateral Export Controls', 12 Apr. 2000, available at URL <http://govt-aff.senate.gov/041200_holum.htm>.

The challenge of nuclear weapons

The use of nuclear weapons by the USA during the closing stages of World War II was an event without parallel in the history of mankind. It was a revolutionary leap in the technological power of mass destruction that proved ultimately to be decisive in ending the war. It was therefore natural that other nations with either serious security concerns or power-projection ambitions would sooner or later find ways to acquire nuclear weapon technology. The Soviet Union, the prime adversary at the time, was quick to develop its own nuclear arsenal. By 1951 US policy makers had decided that an unbridled conventional arms race would have meant 'seeking military safety at the cost of economic disaster'. Scaling up the nuclear arsenal was therefore the choice to 'bring peace power at bearable cost'.[109]

for its security sees technology denial as a hostile or unfriendly act. The supplier countries, in turn, are unlikely to have much sympathy for such concerns so long as international politics and balance-of-power considerations lead them to see their own interests as best protected by technology transfer restrictions of their own devising.

This was the point of no return in the history of nuclear weapon proliferation—the beginning of the all-out superpower effort to build nuclear weapons by the thousands that became the fountainhead of 'nuclear proliferation'. It is open to speculation whether, if such a decision had not been taken by the Soviet Union and the USA, the world might perhaps be very different—with minimal or no proliferation. Atomic technology might then have progressed only for peaceful energy-related purposes, rather than threatening lives on a mass scale all over the world.

The US strategy proved to be correct in terms of keeping the peace and ultimately 'winning' the cold war, but it also had cost implications for the USA. Between 1940, when the USA began work on the atomic bomb, and 1995 the USA spent almost $4 trillion (in 1995 dollars) or about one-third of total US military expenditure during that period, on developing its nuclear arsenal.[110] The cost to the Soviet

---

[109] Schwartz, S. (ed.), Nuclear Weapons Cost Study Project Committee, 'Four trillion dollars and counting', *Bulletin of the Atomic Scientists*, vol. 51, no. 6 (Nov./Dec. 1995), pp. 32–52.

[110] Schwartz (note 109).

Union must have been comparable, which certainly contributed to its ultimate downfall. The cost of acquiring nuclear arsenals for the other nuclear weapon states would have been small in comparison but certainly not negligible in actual terms. If all associated infrastructure expenditure, and the cost of the decommissioning of thousands of warheads and production infrastructures, are added, at least $9 trillion was probably spent during the 20th century on nuclear weapon technologies. This was a huge drain on the world's resources that could have served humanity better if spent in a more productive manner.

There appears to be general agreement that nuclear weapons and the other WMD pose a serious hazard to humanity and that the use of nuclear weapons, except under extreme circumstances, ought to be internationally unacceptable. A universal solution to contain and reverse the proliferation of nuclear weapons could not emerge in the 1950s because the world was recovering from World War II and was also ideologically divided between two adversarial superpowers. Furthermore, the nuclear technological revolution had generated interest that was not as yet balanced by a proper understanding of the consequences of such rapid and phenomenal technological advances.

At the peak of the cold war in the 1970s and early 1980s, both the superpowers and their allies produced tens of thousands of NBC weapons and perfected the various aircraft and missiles needed for their delivery.[111] In terms of international security and stability, this was clearly the most active phase of nuclear weapon and missile proliferation, albeit 'vertical' in nature. In retrospect, this phase was at the root of the rise in the number of nuclear weapon aspirants and the consequent horizontal proliferation during the late 1970s and early 1980s.

Notwithstanding the emergence of the three de facto NWS, a combination of the IAEA, the NPT and the NSG did succeed in persuading some countries to give up their nuclear weapon ambitions and contained 'horizontal' proliferation to the minimum, unavoidable level. On the positive side, since the end of the cold war there has been a decline in the total number of weapons and missiles in the world and the trend has now started to shift in three significant ways.

First, there is a clear recognition that large nuclear warhead stockpiles are not only unusable but also potentially dangerous and eco-

---

[111] See, e.g., the Internet site of the Federation of American Scientists at URL <http://www.fas.org>.

nomically disastrous. Hence, the control of vertical proliferation has already commenced through coordinated nuclear disarmament, with little chance in future of any country aspiring to the production of huge arsenals of nuclear weapons. Second, there appears to be an emerging international consensus that the level of unacceptability of CBW is even higher than that of nuclear weapons. Unlike the NPT, which legitimized five NWS in 1968, the BTWC and the CWC have evolved as non-discriminatory conventions with more support from more nations than ever before. Third, there are grounds to argue (and some states appear to recognize) that continued ownership of nuclear missiles by responsible, stable nations may be much less hazardous than the dangers of such capabilities spreading to fundamentalist, terrorist organizations or to unreliable, unstable or 'rogue' states with a proven record of irresponsible behaviour.

At the start of the 21st century, there is thus a wider appreciation and near-universal agreement on the urgent need to refocus on preventing further proliferation of WMD technologies, particularly to prevent their possible misuse by rogue states and terrorist elements against civil society. This is a clear reorientation of threat since the cold war years, when perceptions were governed by the fear of an all-out nuclear war. The likelihood of full-scale war between stable, democratic, progressive nations now seems remote. Barring the remaining regional tensions linked to border disputes or religious or ethnic tensions, most conflicts are based on economic, environmental resource- and strategic resource-related issues, resolution of which is unlikely to demand the use of WMD. This has created a unique opportunity to review the whole subject of non-proliferation and complete and universal disarmament, at least for CBW and radiological WMD[112] and perhaps even to achieve deep cuts in the number of nuclear weapons.

With regard to nuclear weapons, however, the continued existence of large arsenals in the existing NWS, and their continued reliance on

---

[112] Biological, chemical and radiological weapons are not capable of mass destruction in the same sense as nuclear weapons are, but they are potentially dangerous to human life and ecosystems and have the potential to cause mass disruption and panic. The development and production infrastructures required for these weapons are relatively simple and cheaper than for nuclear weapons. Verification technologies and processes are thus not only very different but also technically very challenging. Recognizing these important differences between CBW and nuclear weapons is the key to adopting appropriate control strategies, which in the first instance must stay focused on complete elimination and in the second instance may accept the possibility of a slower, incremental disarmament of nuclear weapons.

nuclear weapons as the ultimate deterrent, will provide incentives to other states to acquire nuclear weapons to claim equivalent nuclear advantage and privileges. This is detrimental for the curbing of nuclear weapon proliferation and for the future momentum towards eventual nuclear disarmament. However, it can be argued convincingly that, in an increasingly violent world, with so many conflicting and asymmetric forces in play, nuclear deterrence will continue to have a major role in maintaining peace and stability in the near future. Hence, the challenge will be to identify case-by-case motivational factors and create appropriate incentives for additional nuclear weapon aspirants to give up such goals. This can only be done if the justifiable security concerns of all aspirant states are addressed honestly.

Nuclear deterrence and nuclear doctrines

One of the most powerful counterforces to progress in nuclear arms control and disarmament is the unparalleled deterrence value that nuclear weapons provide. As things stand today, the P5 countries have no clear danger of an imminent war situation but still continue to rely on nuclear weapons to deter possible threats to their security from unpredictable elements—and perhaps to balance out any new strategic threats from possible future international conflicts or realignments. For the three de facto NWS, the issue of deterrence is more current because of the regional situations that could easily erupt without a nuclear deterrent.

There is also a perception that a nuclear-weapon capability provides some leverage for resisting coercion by extra-regional major powers. This is a major factor for nuclear-weapon aspirants such as Iran and North Korea, which see deterrence value in terms of freedom to exercise the right to self-defence, as enshrined in Article 51 of the 1945 UN Charter,[113] although they have forsworn the possession of nuclear weapons by ratifying the NPT as NNWS. Their perception of deterrence comes perhaps from the feeling of having more bargaining power within their respective regions as well as against possible coercion by more powerful nations. Unless this deterrent value attached to nuclear weapons can be reduced or replaced by other effective politi-

---

[113] Article 51 of Chapter VII of the Charter of the United Nations is available at URL <http://www.un.org/aboutun/charter/chapter7.htm>.

cal means, the prospects for deep nuclear disarmament remain poor in the near future.

It should be obvious that nuclear weapons have little or no deterrence value against a non-state terrorist organization. Nor do rogue or terrorist entities seek WMD technology for any deterrence benefits. Their purpose is to gain an asymmetric offensive capability. They therefore find CBW equally useful, if not more so. It must also be recognized that any state seeking such asymmetric offensive advantages from WMD technologies risks becoming categorized as a rogue state and, given the robust international response to rogue behaviour in the past few years, it is perhaps less likely that new rogue states will emerge. The Iraq case could serve as a serious warning against future adventurism against international norms.

The experiences of the cold war years have helped to embody the notion that nuclear weapons are weapons of deterrence, not for war fighting, and that a certain minimum capability is needed for 'credible deterrence'. However, how much is enough is always dependent on how much is known about the adversary. The tendency is always to insure against the worst-case scenario.

Another technology-related factor is the perception that enhanced levels of deterrence can be obtained from the possession of thermonuclear weapons, which are far more effective for destroying enemy assets. Fortunately, the technological complexity of thermonuclear weapons is high and currently beyond the reach of adventurist states or groups. Similar arguments would be valid for advanced technologies and the new generation of nuclear weapons. However, technology diffusion makes this a time-limited advantage as more effective technology becomes accessible to a larger number of users, including some with questionable credentials or unpredictable intentions. Arms control and technology controls help to stretch this advantage so that there is time for more advanced R&D to maintain safety margins by providing further options to help prevent the misuse of available technologies.

In earlier times, the pursuit of military technology was essentially aimed at winning wars and 'deterring the enemy's deterrence'. This was primarily an offensive policy. In the 21st century, the focus of the pursuit of technology is more on defensive capabilities and economic gain—more for protection and peaceful competition. Technology will continue to provide strategic advantage in global or regional balance-of-power equations. However, in other areas there will be a significant change in the way modern society perceives defence technology. This is a major shift that should be utilized for positive movement towards complete WMD disarmament.

Nuclear doctrines are based on a nation's own perceptions of the minimum deterrence required to serve its security needs. Enunciation of a nuclear doctrine is an essential component of acquiring deterrence because the deterrence value depends on the projection of a state's capabilities and resolve. A no-first-use doctrine is obviously a defensive posture, requiring a minimum credible deterrent based on the known capacity to retaliate in order to inflict unacceptable damage on the aggressor who first uses WMD. However, reality may not allow for such a 'black and white' or clear-cut distinction. One positive aspect of no-first-use is that it clearly supports non-proliferation and disarmament objectives by reducing the importance of nuclear weapons to a purely defensive role.

It is arguable that nuclear weapons are safer in the hands of nations that are otherwise militarily capable and economically stable, because such nations are more likely to attach only defensive value to nuclear weapons. For an otherwise weak and unstable nation, by contrast, nuclear weapons provide asymmetric advantages in an offensive mode, and such nations may use the threat of nuclear weapons or even the weapons themselves in pursuit of gains that could not otherwise be achieved. This argument, taken to its logical conclusion, explains why nuclear weapons or other WMD would be most dangerous of all in the hands of non-state extremist groups. The arms control community should focus more on irresponsible ownership of the most destructive weapons or sensitive technologies in future, and discriminate more between the safe and unsafe uses of technology. This rationale is discussed further in chapter 5 of this report as a basis for a possible new approach to technology controls.

Returning to the issue of no-first-use, an unqualified no-first-use policy is also likely to reduce the deterrence value of nuclear weapons against the use of CBW and thus encourage the proliferation of these weapons. This realization has brought about a doctrinal change in no-first-use policies to include the use or threat of use of any WMD as an adequate reason to retaliate with nuclear weapons. Some NNWS may perceive this as a violation of the NPT assurances regarding the use of nuclear weapons against NNWS. However, given the near-universal agreement on the bans embodied in the BTWC and the CWC—a condition that did not exist when the NPT was configured—and the emerging tendency of some states to use WMD capabilities for bargaining on regional issues, a qualified no-first-use doctrine seems justified and could form a universal standard among all NWS. It is certainly a way to reduce the dangerous potential of nuclear weapons without affecting their deterrence value. It is important to recognize that universal nuclear disarmament would be feasible only after a comprehensive no-use agreement has been achieved among all the nuclear weapon-capable states and that no-first-use is the first logical step towards no-use.

In the 21st century, the forces responsible for nuclear proliferation are more regional than global, and several NWS now see deterrence as operating in the tactical context. The doctrines and the nuclear postures of the future may be moulded dynamically to meet regional security situations. It is possible that in the future there may be a two-tiered nuclear world, in which the 'nuclear West' led by the USA enjoys a more stable nuclear deterrence and the 'nuclear rest' must balance a minimum credible deterrence in tune with changing security dynamics.

An additional factor that will significantly influence the perceptions of deterrence among the industrially advanced nations is the introduction of ballistic missile defence (BMD) technologies and the weaponization of outer space.[114] There have been fears that US BMD progress could trigger the reactivation of multiple independently targetable re-entry vehicle (MIRV) warhead technology by China and Russia, accompanied by a qualitative enhancement of their respective nuclear arsenals.[115] This could create chain reactions in Europe, South Asia and elsewhere. These various aspects of nuclear deterrence have a major bearing on the future of non-proliferation and disarmament.

Another important issue for progress on disarmament is the subject of the inspection, monitoring and verification of treaty compliance

---

[114] See also section IV of this chapter and section II of chapter 4 in this volume.

[115] At the same time, several countries, including India, have programmes under way to develop missile defence systems. South Korea has indicated that it is considering the creation of an independent missile defence capability. Other countries are modifying existing air defence systems to give them some anti-missile capability. In Russia the S-400 air defence system has been touted as being at least as capable as the PAC-3. There are also several missile defence development programmes under way involving significant international cooperation with the USA. Kile, S. N., 'Ballistic missile defence', *SIPRI Yearbook 2004* (note 56), p. 653.

and the non-proliferation obligations of all states parties in the various arms control regimes. With the availability of increasingly sophisticated technologies for such purposes and the valuable experience gained through years of practice, it should be possible to establish reliable verification systems. However, some of the finer points of dual-use technologies and the capacity of a sovereign state to maintain secrecy pose serious difficulties for verification. In the changing dynamics of arms control and disarmament, influencing intentions through political means may emerge as a more effective and lower-cost method of control. Wider international cooperation will therefore become essential for the future success of arms control.

A simulation case study examining how nuclear deterrence could work in a possible future crisis between China and the USA over the status of Taiwan brings out some interesting lessons.[116] While the real challenge would be to deter effectively the challenger's deterrence, the stakes involved in a regional conflict would be much higher for the regional power than for a distant superpower, and this could seriously compromise what otherwise might constitute a robust deterrence. The study thus concludes that 'deterrence is inherently unreliable'. Instead, it advocates a system of BMD and argues against deep nuclear disarmament in order to maintain full flexibility for unknown future challenges. While this may not be a positive indicator for the future of disarmament, the case study is useful for elucidating the possible weakness of nuclear deterrence in the present era, unless backed by defensive technology capabilities, suitable international cooperation packages and an overall international consensus on no-first-use of nuclear weapons or any other WMD.

II. The Non-Proliferation Treaty and the Comprehensive Nuclear Test-Ban Treaty

Nuclear science arrived almost ahead of its time with huge potential for both peaceful and military applications. Initially, recognition of its potential spurred cooperative efforts to encourage peaceful applications and prevent dangerous uses. The Pugwash Conferences on Science and World Affairs, first held in 1957, was initiated by Albert Einstein and Bertrand Russell to enable scientists from all over the

---

[116] Payne, K. B., *The Fallacies of Cold War Deterrence and a New Direction* (University Press of Kentucky: Lexington, Ky., 2001).

world to unite to address the threat to humanity of the advent of thermonuclear weapons.[117] The IAEA was established in 1957 to encourage and coordinate R&D on atomic energy for peaceful applications and to administer safeguards to ensure that such cooperation does not aid military objectives.[118] However, development of nuclear weapons progressed unabated, along with numerous nuclear weapon tests to establish their military effectiveness. Since the early 1950s, advocates of nuclear disarmament have regarded a ban on testing as an effective way to limit nuclear weapon proliferation and allow movement towards nuclear disarmament—even before the NPT was concluded. The 1963 Partial Test Ban Treaty (PTBT)[119] banned atmospheric nuclear tests but served only to drive the testing underground and failed to contain nuclear weapon proliferation in either its horizontal or vertical dimensions. The NPT entered into force on 5 March 1970 with five NWS as core parties and another set of NNWS, which has now grown to a total of 189 parties. However, between 1970 and the 1995 NPT Review and Extension Conference the total number of nuclear weapons increased dramatically and major advances in technological sophistication were achieved. The NPT thus failed completely in terms of the Article VI commitments to move towards general and complete nuclear weapon disarmament.[120]

The 1998 nuclear tests in India and Pakistan are often cited as another failure of the NPT. However, it can also be argued that at the time of the 1995 NPT Review and Extension Conference India, Israel and Pakistan were already known to be nuclear weapon-capable countries and the failure of the NPT was in its refusal to even notionally accept the reality of the three non-signatories with their existing nuclear options. Even today, the NPT continues with the unrealistic format of not accepting any new NWS other than the P5. There may

---

[117] The Pugwash Conferences is a forum to bring together influential scholars from around the world to discuss the threat posed to civilization by the advent of thermonuclear weapons. See URL <http://www.pugwash.org>.

[118] See note 8.

[119] The 1963 Treaty Banning Nuclear Weapon Tests in the Atmosphere, in Outer Space and Under Water (Partial Test Ban Treaty, PTBT), which entered into force on 10 Oct. 1963, prohibits the carrying out of any nuclear weapon test explosion or any other nuclear explosion: (*a*) in the atmosphere, beyond its limits, including outer space, or under water, including territorial waters or high seas; and (*b*) in any other environment if such explosion causes radioactive debris to be present outside the territorial limits of the state under whose jurisdiction or control the explosion is conducted. The text is available at URL <http://www.unog.ch/frames/disarm/distreat/part_ban.htm>.

[120] Goldblat (note 105), pp. 106–107.

be yet more nuclear weapon-capable states in the future, unless and until the tide is turned in favour of near-universal nuclear disarmament. Such developments would be fatal for the credibility of the NPT and provide a powerful impetus to consider some pragmatic amendments.

The primary focus of the NPT, assisted ably by the Zangger Committee[121] and the NSG, has been to prevent nuclear weapon capabilities from spreading to other states. Over the years, the IAEA has gathered a stature and teeth that few other international organizations can boast of. The 2000 NPT Review Conference sought to strengthen verification of treaty compliance through an additional protocol for compliance verification and nuclear safeguards with short-notice inspections and environmental sampling both from non-proliferation and safeguards perspectives.[122] The nuclear non-proliferation movement has a robust organizational structure to prohibit further proliferation. When Belarus, Kazakhstan and Ukraine gave up their inherited nuclear arsenals and acceded to the NPT as NNWS, this was indeed an important success for the NPT.[123] However, for the ultimate success of the NPT, NWS have a responsibility to serve as role models for non-proliferation in every sense of the overall spirit of the NPT, including their Article VI commitments.

In the context of progress towards complete nuclear weapon disarmament, the CTBT and a possible fissile material cut-off treaty[124] would—as argued above—be valuable disarmament tools. While the CTBT could limit the qualitative advancement of nuclear weapon development, the FMCT, if configured in a globally acceptable man-

---

[121] Established in 1971 and named after its first chairman, the Zangger Committee is a group of nuclear supplier countries that meets informally twice a year to coordinate export controls on nuclear materials. For the participants see table 3.1 in this volume.

[122] See Simpson, J., 'The 2000 NPT Review Conference', *SIPRI Yearbook 2001: Armaments, Disarmament and International Security* (Oxford University Press: Oxford, 2001), pp. 487–502.

[123] The 1991 Soviet/Russian–US Treaty on the Reduction and Limitation of Strategic Offensive Arms (START I Treaty), which entered into force on 5 Dec. 1994, obliges the parties to make phased reductions in their offensive strategic nuclear forces over a 7-year period. It sets numerical limits on deployed strategic nuclear delivery vehicles, intercontinental ballistic missiles, submarine-launched ballistic missiles and heavy bombers—and the nuclear warheads they carry. In the 1992 Protocol to Facilitate the Implementation of the START Treaty (Lisbon Protocol), which entered into force on 5 Dec. 1994, Belarus, Kazakhstan and Ukraine assumed the obligations of the former Soviet Union under the treaty. They pledged to eliminate all the former Soviet strategic nuclear weapons on their territories within the 7-year reduction period and to join the NPT as NNWS in the shortest possible time.

[124] See note 101.

ner, could limit quantitative proliferation in both its horizontal and vertical dimensions. As a traditional champion of nuclear disarmament, India enthusiastically supported the PTBT in 1963 as a positive step towards serious non-proliferation efforts. It was only after the end of the cold war that the issue of a test ban resurfaced because a new window of opportunity was available to address nuclear proliferation and disarmament. In 1993 India co-sponsored the idea of a CTBT along with the USA, once again in the hope of achieving some progress towards disarmament, only to find that there was no real commitment to universal nuclear disarmament and that the CTBT was being used to close all options for the rest of the world, irrespective of regional security requirements. Since much has been published on the CTBT negotiations and its present uncertain status, this report addresses only some of the relevant issues in the context of its non-proliferation objectives, leading to the eventual possibility of total and universal nuclear disarmament. The approach taken here aims to bring out the point of view of a country such as India that sees itself as having an excellent record of nuclear restraint and responsible nuclear behaviour, yet continues to be perceived as a country of nuclear proliferation concern.

The reasons behind US-led initiatives on the CTBT and an FMCT in the early 1990s can be understood more easily with the benefit of hindsight. Having reached technological sophistication long before the other NWS, the USA stood to gain the most from a moratorium on nuclear testing. The CTBT negotiations provided an opportunity to freeze technological superiority vis-à-vis other NWS as well as to cap the threshold nuclear nations—India, Israel and Pakistan—at the time. Israel, of course, by drawing *inter alia* on French and US technology had already acquired the expertise as well as the minimum necessary stockpile for its security. Pakistan, with access to assistance from China, perhaps had no technological need to conduct nuclear tests of its own. Hence, the CTBT would have had the selective effect of closing India's nuclear options while protecting the interests of the NWS, which could safely maintain their stockpiles using procedures permitted under the treaty.[125] From India's point of view, the CTBT was thus emerging as another discriminatory control regime rather than a valuable disarmament tool and India therefore rejected the CTBT in this form, even at the cost of being isolated internationally.

[125] See note 26.

Since the beginning of the nuclear age India has championed the cause of universal nuclear disarmament and rejected any proposal falling short of this. However, Indian disillusionment with the concept of universal nuclear disarmament began in the 1980s when China, its north-eastern neighbour, started equipping its western neighbour Pakistan with nuclear weapon technology, and when US-led international non-proliferation regimes failed to react strongly to this because of other political priorities. In June 1988, Indian Prime Minister Rajiv Gandhi enunciated his action plan for a nuclear weapon-free world, in a desperate attempt to attract world attention at the special session on disarmament of the UN General Assembly.[126] Since the action plan failed to achieve its objectives and there was physical evidence that Pakistan was acquiring a nuclear weapon capability, India had no choice but to accelerate its own nuclear weaponization programme. As mentioned above, India co-sponsored the early negotiations on the CTBT in 1993. However, throughout the negotiating stages at the Conference on Disarmament (CD) in Geneva, it became clear that the CTBT was going against India's national interest. India therefore decided not to sign the CTBT—eventually acquiring a bad name for blocking the ratification of the treaty. The Indian Ambassador to the CD forcefully articulated the main reasons for the decision: (*a*) the CTBT, in the form being finalized, seriously compromised India's major national security concerns; (*b*) the entry-into-force clause of the CTBT made Indian ratification of the treaty necessary for its implementation;[127] (*c*) the NWS had failed to demonstrate any commitment to eventual total nuclear disarmament; and (*d*) the CTBT was to be based on a verification regime as discriminatory as the NPT, which India had steadfastly refused to join.[128] Nevertheless, India was labelled as the 'treaty spoiler' although subsequently the US Congress did not ratify the CTBT. The Bush Administration has shown even

---

[126] Gandhi, R., 'Action plan for a nuclear weapon-free and non-violent world order', *Selected Speeches and Writings 1988* (Government of India, Ministry of Information and Broadcasting: New Delhi, 1989), pp. 331–41.

[127] India believes this to contravene the 1969 Vienna Convention on the Law of Treaties, which forbids compelling a sovereign state to become a party to a treaty that is manifestly not in its national interest.

[128] Ghose, A., 'Negotiating the CTBT: India's security concerns and nuclear disarmament', *Journal of International Affairs*, vol. 51, no. 1 (summer 1997), pp. 239–61. A version of this paper is also available on the Internet site of the Indian Embassy, Washington, DC, at URL <http://www.indianembassy.org/policy/CTBT/ctbt_ghose.htm>.

less enthusiasm about its coming in to force and the fate of the CTBT remains uncertain.

Negotiations are yet to begin on an FMCT but here also there are many contentious issues—mainly connected to existing stocks, and to relations between the nuclear haves and have-nots as well as between the nuclear weapon-capable states. India and Pakistan, having crossed the nuclear weapon threshold in May 1998, might take a different view if an FMCT were to facilitate a formal acceptance of their nuclear weapon status. Israel, with its unquestioned, albeit officially undeclared, nuclear weapon status might also support an FMCT, if it would help to freeze the Middle East nuclear balance in its favour. It is clear that an FMCT, if managed skilfully, could contribute significantly to international nuclear non-proliferation and disarmament efforts.

The bilateral agreements between Russia and the USA on non-production of highly enriched uranium (HEU) and plutonium are important steps that need to be made more transparent in order to provide momentum for a global FMCT.[129] In terms of prohibiting the misuse of fissile material by rogue or terrorist elements, a comprehensive, transparent and universal 'non-production of fissile material' treaty will be vital to safety concerns. Such a treaty, if implemented successfully, would certainly end vertical proliferation and may well set the stage for future global cooperation on issues of nuclear non-proliferation and disarmament, in keeping with the commitments of the 1995 NPT Review and Extension Conference.

As regards the status of the NPT and nuclear non-proliferation performance, in addition to the five recognized NWS, today there are three more de facto NWS which are neither formally recognized as NWS nor parties to the NPT. North Korea may be on the threshold of declaring itself a nuclear weapon state and has opted out of the NPT—the first nation to do so. Although Iraq has now been fully contained there are concerns that Iran may be developing nuclear weapon technology, although it may be years away from achieving any nuclear weapon capability. Germany, Japan and South Korea are other countries that might have the technological capacity to rapidly develop nuclear weapons. South Africa destroyed its substantial

---

[129] See Bunn, M. and Holdren, J. P., 'Managing military uranium and plutonium in the United States and the former Soviet Union (excerpts)', (n. d.), available on the Internet site of the Russian American Nuclear Security Advisory Council at URL <http://www.ransac.org/Issues/bunnholdren1.html>.

nuclear weapon capabilities because of concern about a nuclear arms race in Africa and also over the issue of the command and control capabilities of the new government. Argentina and Brazil gave up their nuclear aspirations as a result of the successful 'carrot and stick' policy of the USA.

Overall, the world nuclear scene appears to be much more stable than in the final decade of the 20th century, when so many unprecedented changes occurred in such quick succession.[130] The proliferation problem is no longer global or generic but regional and specific. The next few decades could be managed effectively to further enhance this stability and take advantage of international willingness for wider and deeper international cooperation in the service of long-term non-proliferation and disarmament objectives.

III. The changing environment for export controls

The end of the cold war removed the rationale for traditional export controls quite suddenly, as the ideological and military threats to the NATO member states dissipated. The position of export controls on the national security agenda was no longer self-evident, and questions emerged about their continued relevance. The disappearance of the bipolar balance between the superpowers also caused the slow breakdown of several delicately poised regional balances, leading to a rise in regional instability. Conflicts emerged in many parts of the world and a number of local clashes were accentuated, particularly in the developing regions. As tensions in many other areas also became heightened, most nations were compelled to revise their security perceptions, leading to subtle changes in political priorities.

It was this changed security perception combined with opportunism that prompted Iraqi dictator Saddam Hussein to invade Kuwait and led to the 1991 Gulf War. It was also this event that brought export controls back into focus, since post-war analysis showed Iraq to be in gross violation of its treaty obligations as a party to the NPT. It had not only successfully pursued the development of nuclear weapons but also stockpiled CBW agents for use. The Gulf War also brought into focus the fact that Iraq's requirement for delivery systems was well within the 300-km limit set by the MTCR and that this weapon delivery range could be attempted with a mega-cannon delivery sys-

[130] Cirincione (note 98).

tem, which Iraq was pursuing. However, much of what Iraq had acquired, directly or indirectly, was provided by the various supply-side countries that supported export control mechanisms. Obviously, there must have been many gaps in the prevailing export regulations for them to permit such significant trade in prohibited products and technologies. It may be relevant to note that the same Iraqi regime had been supported by Western nations in its 1980–88 war against Iran—itself considered a close ally of the USA during the 1960s and 1970s. Export control has therefore been very much an 'adversary-oriented' foreign-policy tool that changes with time, and that carries its own mix of short-term advantages and long-term drawbacks.

While arms control originally evolved as an agreement between adversaries or competitors to reduce the cost of arms races, and arms control objectives are thus supported by international treaties with all affected parties as consenting participants, export controls—as noted above—have evolved as agreements between allies to limit the techno-military capabilities of the enemy through technology controls and export regulations on technology transfers. These measures can be employed unilaterally but are more effective if imposed multilaterally. The target of export controls, however, in all events is not a willing participant in the process.

Arms control continues to be relevant in the post-cold war era because it is an essential component of the preventive defence doctrine. It promises to reduce the cost as well as the risk of maintaining huge stockpiles of deadly arsenals that have the potential to cause unacceptable damage, including through accidents. The rationale behind export controls, however, has undergone significant change. The supply-side groups still perceive them as relevant to their overall non-proliferation agenda. However, the demand side has difficulty in understanding the non-proliferation benefits of politically motivated export controls and tends to perceive technology denial or control as detrimental to its legitimate requirements for defence and development. There is also a school of thought which believes that sufficient non-proliferation controls already exist under the BTWC, the CWC and the NPT. More generally, the tendency to continue with dual-use technology denial policies under informal export control arrangements such as the MTCR and the WA, which are largely politically motivated, appears to be in contradiction with post-cold war initiatives to forge cooperation with former non-allies. It also impacts on the sovereign right of nations to pursue technology for both security and economic reasons in the globalized world of the 21st century.

Technology controls are indeed important and necessary, but future control systems need to address the sensitivities of both the demand and supply sides to strike a balance between control and cooperation in international practice. Since the reasons for and parameters of controls are changing, this report argues that future technology controls will require a management-oriented approach rather than a control-dominated policy. The distinction has also been made between arms control and disarmament, which share a common non-proliferation objective, and export controls and technology control, which share common technology-management challenges for the future.

In the years since the end of the cold war the concepts and practices behind *arms control* have raised questions over how much of the old practice is relevant or even legitimate.

The USA, which has historically been the prime mover behind all arms control and disarmament initiatives, is now critical of traditional arms control arrangements, seeing them as unbalanced, unenforceable and irrelevant to real contemporary threats; economically intrusive and expensive; and even as an obstruction to new and better measures to combat real threats.[131] With the paradigm shift in the security environment around the world, arms control not only needs to be revised at the conceptual level but also needs to be made more pragmatic and packaged to suit the specific demands of the future.[132]

The rationale for *export controls* has also changed. The main threat to Western developed nations is now seen as coming from terrorist organizations and their state supporters, which are often identified as rogue states. Therefore, the architects of arms control must take a serious look at how to restructure the existing regimes and how best to address the most immediate need—to ensure that potentially dangerous WMD technologies do not get into the wrong hands.

Technology is impartial and the effects of globalization, driven by market forces, now enable much smoother technology transfers

---

[131] 'Multilateralism and arms control are both under threat, principally from the rise to power in the United States of a neo-conservative analysis that portrays international treaties and alliances as "vitality sapping, virility constraining and option closing" encumbrances for a country that believes it has the military and economic dominance . . . to do as it pleases.' Johnson, R., 'Incentives, obligations and enforcement: does the NPT meet its state parties' needs?', *Disarmament Diplomacy*, no. 70 (Apr./May 2003), pp. 3–10.

[132] Anthony, I., 'Arms control in the new security environment', *SIPRI Yearbook 2003* (note 6), pp. 563–76.

through the faster movement of products and people across international boundaries. Human ingenuity has been the main engine of the revolution in information—more so than with traditional engineering innovations—and this has changed the balance of knowledge holdings. The advanced industrial countries are, for the first time, seeking expertise and knowledge from some progressive developing nations. Interestingly, while the advanced industrial nations have continued to deny technology to most developing nations, the developing nations, for their part, are neither organized nor inclined to deny their knowledge and expertise to the advanced industrial nations. In the changing international system, it is important to appreciate the other side's point of view and also to recognize the levelling effects of the diffusion of advanced technology.

For most of the technology supplier states, the changes in export control priorities are now more focused on economic competition than on the predominant security preoccupations of the cold war years. The definition of an enemy has become blurred, and even the identity of adversaries and friends has become subject to changing techno-economic priorities and changing times. It has been calculated that multinational companies, with transnational suppliers, production centres and financial support, are responsible for more than half of the world's industrial output.[133] Government agencies are no longer able to exert the same level of control over technological development or technology acquisition. Economic compulsions have created a new culture of modular designs for using off-the-shelf components to save on R&D costs, and major industrial concerns regularly outsource many of their assembly and manufacturing jobs to more cost-effective regions of the world, many of which are in the developing nations.

Recent revelations about the WMD capabilities of Iran, Iraq and North Korea suggest that these states owe a great deal to the export decisions made deliberately or inadvertently by supplier states such as

---

[133] See Vernon, R., *In the Hurricane's Eye: The Troubled Prospects of Multinational Enterprises* (Harvard University Press: Cambridge, Mass., 1998). It has also been calculated that only 5 of the world's largest spenders on R&D are governments (the single greatest spender being Ford Motors) and that some 25 of the world's 100 largest 'economies' are corporate entities rather than states. See van Tulder, R. K., 'The power of core companies', PriceWaterhouseCoopers, *European Business Forum*, 2002, excerpt available at URL <http://www.pwcglobal.com/Extweb/NewCoAtWork.nsf/docid/11CB933A2735DD6685256C9500551693>.

China, France, Germany, Pakistan, Russia and the USA.[134] Even if export controls were to function ideally, making acquisition of WMD technologies more expensive and time-consuming, export controls alone cannot and were never designed to end proliferation. Hence, over time several other mechanisms have evolved to assist the non-proliferation objectives of export controls. These include multilateral sanctions and incentives, bilateral arms control agreements, confidence-building measures and, when all else fails, counter-proliferation action.

The technologies, products and services that have been subjected to these controls are categorized in different groups depending on their different applications. There appears to be almost unanimous consensus about controlling WMD. However, even within this category, there are large variations in technological complexity, destruction potential, infrastructure requirements, and the possible verification tools for monitoring and control. All of these factors have different acceptability or unacceptability quotients and, often, considerations of prestige attached to them. For instance, if the same national security yardsticks are to be applied universally, countries with indigenous technological capabilities and legitimate security reasons cannot be faulted for seeking nuclear and missile technologies just because they are out of synchronization with the timetables of certain treaties or the prevailing priorities of the five NWS. Some NNWS are beginning to raise such questions. At the 2003 Non-Proliferation Treaty Preparatory Committee meeting, the continued insistence of the P5 nations on their right to indefinite dependence on nuclear weapons prompted Sweden to question the rationale behind this insistence by France and the UK, and to ask why a country such as Sweden should not be governed by the same rationale.[135]

There is a general international convergence of views on the need for arms control and WMD disarmament, especially for CBW, as is evident from the broad agreement reached on the unacceptability of these weapons. The category of major conventional weapon systems, however, represents a different challenge because it offers an attractive market for the defence industry and there are no universal institutionalized controls in place. Generally, transfers of these weapons are

---

[134] Beck, M. and Gahlaut, S., 'Creating a new multilateral export control regime', *Arms Control Today*, vol. 33, no. 3 (Apr. 2003), pp. 12–18.

[135] Johnson, R., 'Rogue and rhetoric: the 2003 NPT PrepCom slides backwards', *Disarmament Diplomacy*, no. 71 (June/July 2003), pp. 3–23.

governed more by political parameters weighed against commercial interest. If the P5 is the smallest and tightest control group, even here there is disagreement on the issue of high-value defence exports, as was evident when China ceased to cooperate on export controls after the US decision to sell F-16 combat aircraft to Taiwan in 1992.[136] Differences in perception on the issue of desired controls on light weapons, services and dual-use technologies are even wider within the larger group of nations forming the various multilateral export control regimes. In addition, the numbers of suppliers for some of these items have increased significantly and not all have the same priorities or are participants in the same control regimes.

Along with WMD proliferation, another area of widespread agreement regarding technology and export controls is the identification of the common enemy, comprising irresponsible rogue states and faceless non-state players that use terrorism, often driven by fundamentalism, to oppose the authority and power of the internationally established system. Mechanisms to prevent or reduce these high-priority dangers through preventive security actions must now be at the heart of export controls. However, because this phenomenon is relatively new, there are as yet no clear definitions of what exactly constitutes the enemy. Most nations are following the superpower focus on transnational terrorism of the al-Qaeda type, while some others, such as India, are fighting their own battle against other equally dangerous fundamentalist groups that have carefully stayed away from the US radar screen but are more active and visible in their regions of interest. While rogue states and terrorist groups identified by the USA are subjected to strict controls and sanctions, some other regional terrorist groups are suspected of continuing to receive substantial support from states such as Pakistan, recognized as an important partner in the US-led fight against terrorism. Export controls and proliferation concerns are placed on the back burner in such cases, although the potential dangers are obvious. Terrorism cannot be contained unless it is confronted globally—using a long-term focus and an unbiased, comprehensive definition. Terrorist organizations have acquired a chameleon-like presence—camouflaged under different names and

---

[136] Anthony, I. et al., 'Arms production and arms trade', *SIPRI Yearbook 1993: World Armaments and Disarmament* (Oxford University Press: Oxford, 1993), p. 421.

identities and operating globally without fixed structures or infrastructures.[137]

The pre-emptive attack on Iraq by the US-led coalition, without a UN mandate or even the support of some of the USA's longstanding allies, has signalled a new era in the enforcement and control of technology denial where the perception of the sole superpower is supreme, irrespective of international regulations or multilateral agreements. In the process, the credibility of UN systems of inspection and verification has come under serious question. Given the complexity of modern technology, the future challenges for inspections and verification are likely to be far more demanding. Obviously, the control regimes of the future will have to contend with this new reality and re-orient themselves to remain relevant. 'Future multilateral export control regimes will have to broaden the scope, increase the flexibility, simplify the implementation methods and enhance the speed of their achievements.'[138] However, the moot question is how broad the scope of such controls may become without risking the dilution of the core aims or becoming too expensive or too generalized to be successful. The changing environment calls for a total rethink of the conceptual issues around export controls and technology management, as well as of their enforcement.

IV. Missile proliferation, the Missile Technology Control Regime and ballistic missile defence

As ideal delivery vehicles for nuclear warheads, missiles represent a special class of force multipliers. The history of long-range rocket science is well documented and the technology requirements for increased ranges and more sophisticated missiles are fairly well understood. Early advances in technology and escalation in numbers were, again, the product of the cold war years. At the peak of the cold war, the Soviet Union and the USA had enormous missile arsenals deployed against each other at alert status. During the early 1980s, the discovery of the Argentinian–Egyptian–Iraqi Condor II project and the realization that a number of other developing countries possessed,

---

[137] Volkov, V., 'The resources and tactics of terrorism: a view from Russia', eds Bailes and Frommelt (note 35), pp. 111–18.

[138] Moodie, M., 'Constraining conventional arms transfers', *The Arms Trade: Problems and Prospects in the Post-cold War World*, in R. E. Harkavy and S. G. Neuman (eds), *Annals of the American Academy of Political and Social Sciences*, vol. 535 (Sep. 1994), pp. 131–45.

or were developing or acquiring, a large variety of missiles, created a major security concern that led the then G7 to configure behind the scenes the MTCR, which was announced in April 1987.[139] This informal export control arrangement is an attempt to stop the proliferation of NBC weapons by preventing access to delivery systems for them. Since 2002, when South Korea participated in the MTCR plenary meeting for the first time, 33 states have been participating in the MTCR. Other nations, such as China and Israel, have also agreed to adhere to the MTCR Guidelines without joining the regime.[140]

Despite mature understanding of the technological issues involved with missile proliferation, the subject of missile technology control has remained too controversial to make possible the creation of a truly inclusive international regime, with sharp differences between the advanced industrialized group and the less developed group of nations. Differences also exist among the supplier group participants about specific details of the control regime—especially where commercial benefits outweigh perceptions of proliferation threats. While there is general agreement that missile proliferation poses a serious threat to international peace and stability, the threat is not uniform, varying from country to country, region to region and, of course, from time to time depending on the dynamics of political alignments.

Missile proliferators can perhaps be grouped in three distinct categories. The original vertical proliferators consisted of the participants in the supplier group, with the Soviet Union and the USA being that era's biggest contributors to proliferation in terms of variety, sophistication and numbers. The Soviet Union (and now Russia) and the USA have never prohibited the sale of short-range missiles and sounding rocket technology, which still continues, to their allies and friends. The second group consists of those countries with sufficient indigenous design capability to develop missiles for their own requirements (albeit with commercially available support in the areas of compo-

---

[139] Cirincione (note 98).

[140] Israel agreed to apply the MTCR Guidelines through its national export control system in Jan. 1992. China committed itself to adhere to the MTCR Guidelines on 21 Feb. 1992. Three other states that do not participate in the MTCR have also agreed to apply the Guidelines through national export control systems: Romania, in Sep. 1992; Slovakia, in Jan. 1994; and Bulgaria, in Mar. 1996. President George H. W. Bush, 'Message to the Congress reporting on the national emergency with respect to export controls', 31 Mar. 1992, URL <http://bushlibrary.tamu.edu/research/papers/1992/92033101.html>; and US Department of State, Bureau of Nonproliferation, 'Fact sheet: Missile Technology Control Regime', 23 Dec. 2003, URL <http://www.state.gov/t/np/rls/fs/27514.htm>.

nents and sub-systems), but which were not party to the closed negotiations on the MTCR for political reasons. This category includes principally China, India, Israel and South Africa. The third category consists of all the other aspirants to missile capability, largely dependent on foreign supply or technology from abroad. In the 21st century there is concern about a fourth category that consists of terrorist organizations and non-state actors which have been able to acquire and use short-range missiles or rockets to great advantage in locations such as Lebanon.

As described above, the MTCR has adapted over time to apply controls based on the end-use of items as well as their technical characteristics. After the events of 11 September 2001 there have been further clarifications to the scope of the Equipment and Technology Annex to harmonize the application of controls.[141] This process of adaptation, while quite understandable in the context of the unpredictable nature of missile threats today, has created a major paradox for MTCR implementation. The MTCR, which started with unambiguous technical specifications, is now a 'catch-all' regime that is concerned more about user qualifications than technology proliferation. While this approach may seem pragmatic in the changed environment of technology diffusion, the MTCR is not equipped to pass judgement on all transactions involving all potential items that can be delivered remotely to pre-selected targets.

It is possible to illustrate this with reference to cruise missiles, which have now emerged as an equally dangerous threat to that posed by ballistic missiles. Cruise missiles are similar to UAVs in that they are powered and guided all the way to their targets, but the former fly much faster. The ready availability of GPS signals has simplified accurate cruise missile guidance, and delivery to a target with an accuracy of 100 metres is possible. Although they have limited payload capability, cruise missiles are cheaper and quicker than ballistic missiles and have the flexibility to be launched from ships or aircraft. A cruise missile flying low and slowly is now of major concern because of its suitability for delivering biological or chemical agents in a controlled and gradual fashion over a wide area.[142]

The technology for basic cruise missiles is relatively simple and more affordable than that for ballistic missiles. While the MTCR may

---

[141] See the MTCR documents (note 16).
[142] Gormley (note 51).

well have to concentrate more on cruise missiles and even UAVs in the future, here again there are major commonalities with general aerospace and guidance technologies that are now fairly widespread and are used in the commercial domain, where technology diffusion will be unavoidable. Once again the conclusion can be drawn that the proliferation focus must now be on the users.

Another observable trend is that in practice the MTCR has helped facilitate missile development and trade among its participants. Following Brazil's entry into the MTCR in 1993, and the related acceptance of its SLV programme, Ukraine also won a US concession that, as a participant, it could retain and develop missiles with a range of up to 500 km. South Korea, which was limited to 180-km range missiles by a bilateral agreement with the USA, secured US approval for 300-km range missiles and has subsequently made a case for 500-km range missile technology for its SLV project.[143]

China indicated its willingness to observe the MTCR Guidelines in 1992 and finalized a bilateral agreement with the USA in 1994. However, China continues to be of proliferation concern because of its continued missile trade with Iran and Pakistan. During the 1990s, Pakistan received more than 30 complete Chinese M-11 missiles with a 280-km range and a 1000-kg payload as well as Chinese assistance for the production of M-11 class missiles.[144] Pakistan's solid-fuel Shaheen missile with a 750-km range, tested in April 1999, is believed to be based on either a Chinese M-9 or M-11 design.[145] The liquid-fuelled, 1500-km range Ghauri II missile, thought to be based on North Korea's No Dong design, was tested in April 1999.[146] Pakistan's missile tests, which took place in rapid succession, add to the speculation that the missiles represent proven designs obtained from established suppliers.

The formal admission of Russia as an MTCR participant has not solved many of the disputes regarding Russian cooperation with India or Iran. Under its national laws the USA has imposed sanctions on one MTCR participating state (Russia) that it judges to have made

[143] Wagner, A., 'S. Korea, U.S. agree on missile guidelines, MTCR membership', *Arms Control Today*, vol. 31, no. 2 (Mar. 2001), available at URL <http://www.armscontrol.org/act/2001_03/southkorea.asp>.

[144] For more on Pakistan's missile capabilities see Kristensen and Kile (note 57), pp. 624–26.

[145] Kristensen and Kile (note 57), p. 625.

[146] Cirincione (note 98), p. 214. See also Wezeman, S. T., 'Suppliers of ballistic missile technology', *SIPRI Yearbook 2004* (note 56), p. 547.

transfers that violate the established guidelines.[147] However, the MTCR as a body has no sanctions available in cases where such transfers take place. The nature of the MTCR, which rests on political cooperation among like-minded states, would make it difficult to develop either a mechanism for judging non-compliance or a set of sanctions to be applied to a participating state considered to have violated the guidelines. This contributes to the general impression that a country with missile or SLV ambitions could expect easier access to previously denied technology and equipment by becoming a participant in the MTCR, as well as more lenient treatment should it utilize various loopholes in the MTCR for its commercial advantage.[148] The application of sanctions to entities in Russia underlines that the real challenge to the MTCR is the dual-use nature of and the rapid advances in missile-related technologies. Overlap with civilian technologies and applications makes commercial exchange unavoidable and excessive control too complex and expensive.

It is argued here that the failure of the effort to control missile technologies based only on technical performance partly reflects advances in knowledge that have made formerly high-end performance commonplace and often commercially available. It is further argued that the decision to expand the scope of the MTCR through an end-use-based element in the guidelines is not only a major contradiction but also a recipe for failure. With advances in cruise missiles and UAVs, there are innumerable possibilities for innovative missile applications. It might even be questioned why the regime profile might not be expanded to include small manned aircraft because, for an operator on a suicide mission with CBW or a crude radiological weapon, even a range of a few hundred kilometres is enough.

Although the MTCR includes a commitment among participating states not to 'undercut' one another by supplying a controlled item to an end-user that has been denied an essentially identical item by a partner,[149] it is a weakness of the regime is that it lacks any strong mutual obligations or clear universal incentives. Again, although

---

[147] By Feb. 1999 the USA had imposed sanctions on 10 Russian entities, 3 of which were said to be cooperating on Iran's Shahab missile programme. Katzman, K., US Congress, Congressional Research Service, *Iran: Arms and Technology Acquisitions*, CRS report to Congress (US Government Printing Office: Washington, DC, 26 Jan. 2001).

[148] See Ahlström (note 17).

[149] While the no-undercut policy operated within the MTCR from its creation, it was not made public until the Stockholm plenary meeting of Oct. 1994.

MTCR participants are free to cooperate on civilian-use technologies, the actual benefits vary greatly depending on actual political judgements and relationships, because there are no defined standards for military cooperation.

A country-by-country examination of the MTCR's successes and failures might identify Argentina, Brazil, Egypt, South Africa, South Korea, Taiwan and Ukraine as successes and India, Iran, Iraq, Israel, North Korea and Pakistan as failures. With the enhanced scope of the MTCR, it is a moot point whether the number of countries now interested in developing or acquiring short-range rockets, missiles and UAVs should be seen as an MTCR dilemma. The actual failures of the MTCR are linked to cases of supply-side participants and adherent nations that continue to export missile and aeronautics technologies by exploiting the ambiguities of range, payload or application standards. China and Russia have made good use of such loopholes. However, since an assessment of missile technology proliferation control is normally presented in terms of countries of concern to the Western group of supplier countries, its performance should be gauged against the intended political objectives.

What has gone wrong with the Missile Technology Control Regime?

The scope of the MTCR has been progressively widened to cover all delivery vehicles except manned aircraft, which are equally capable of WMD delivery but slower and hence easier to defend against. In some regional contexts, however, manned aircraft can be a more serious threat than missiles, particularly if on a suicide mission. This was recognized in the regions concerned even before the lessons of 11 September 2001. Another problem is that linking the 'proliferation concern' definition to 'intentions' has made it specific to the threat perceptions of a specific group of nations. However, the MTCR now has 33 participants (see table 3.1), as well as several nations that adhere to its principles, and not all of these share the same threat perceptions or missile proliferation concerns.

As mentioned above, and in spite of statements to the contrary by participating states, the MTCR is gradually being perceived as a 'missile supermarket arrangement' where, depending on its political–economic relationship with the USA or other major powers, a country can legitimize its pursuit of missile technology by becoming a participant. The commonality of missile components with space technology and the permissibility of national space projects have created unlimited possibilities for MTCR participants to pursue unspecified long-term missile ambitions.

The rapid growth of technologies for various unmanned delivery systems capable of carrying WMD payloads is of concern to MTCR participants. Recognition of the limits of the MTCR was one reason why many (notably European) participant countries pushed for the Hague Code of Conduct.[150] Because it does not forbid possession of ballistic missiles per se, but instead calls for greater restraint and caution, and wider, more transparent, cooperation to prevent their misuse through information sharing, the HCOC is designed to draw the support of a larger number of nations than belong to the MTCR. In addition, the HCOC does not prevent states from benefiting from technology for the peaceful use of outer space. The HCOC can be seen as a new approach to building mutual confidence among missile-holding nations, reducing mistrust and enhancing levels of cooperation.[151]

Missile defence technology and the Missile Technology Control Regime

Investigating the feasibility of defending against missile attack appears to be almost as old as the development of the offensive missiles themselves. In the 1950s individual weapons with some capability against ballistic missiles such as the US Nike–Ajax and Nike–Hercules air defence missiles were being tested and fielded. The Soviet Union is believed to have been the first country to establish a limited missile defence system to protect Moscow in the late 1960s—a system that was upgraded in the late 1970s.[152] Missile defence technology has been evolving in the USA ever since President Reagan's SDI announcement in March 1983.[153] The 1972 Treaty

---

[150] See note 17.
[151] Chuter, A., 'Missile fears spur nonproliferation pact', *Defense News*, 18–24 Nov. 2002, p. 4.
[152] The Federation of American Scientists (FAS) has established a useful inventory of US historical missile defence systems and projects. See FAS, 'Former programs', 13 May 2003, available at URL <http://www.fas.org/spp/starwars/program/complete.htm>. For Soviet missile defence programs see FAS, 'Soviet BMD programs', available at URL <http://www.fas.org/spp/starwars/program/soviet/index.html>.
[153] See note 25.

on the Limitation of Anti-Ballistic Missile Systems (ABM Treaty)[154] between the two superpowers helped somewhat to prevent a missile defence technology race but the USA withdrew from the ABM Treaty in June 2002.[155] It is pursuing plans to operationalize a rudimentary defence system against long-range ballistic missiles in the near future.[156]

US concerns are currently focused on the possibility of a missile attack from a rogue state or an accidental launch against US territory. These are difficult threats to pinpoint without reference to the known characteristics of a potential adversary. The technical capacity to attack the US mainland still lies with China and Russia although, according to intelligence reports, Iran and North Korea could acquire such capabilities in the future.[157]

Lack of confidence in the success of the MTCR and concern about the proliferation of advanced ballistic missile capabilities has prompted the USA to vigorously engage in a multi-layer BMD programme designed encompassing short-, medium- and long-range missiles for defence against ballistic missiles at various stages in their trajectory. Most of the proposed systems are 'hit-to-kill' type systems with demanding technological specifications that are yet to be proven.[158] Apart from the USA, Israel and Russia are countries with known missile defence efforts. France and Italy are developing the Aster air defence missile system, which has a stated role in defending against ballistic missiles. China and India are also believed to have substantial indigenous technology programmes for missile defence, particularly for terminal phase defence. A number of NATO member states (e.g., Germany, Italy and the Netherlands) as well as Australia, Japan, South Korea and Taiwan stand to gain significantly from US missile defence technology. The countries of the Gulf Cooperation

---

[154] The 1972 Treaty on the Limitation of Anti-Ballistic Missile Systems (ABM Treaty) entered into force on 3 Oct. 1972 but is not in force as of 13 June 2002. For the text of the treaty see *UN Treaty Series*, vol. 729 (1970).

[155] For more on the US decision to withdraw from the treaty see Kile, S. N., 'Ballistic missile defence and nuclear arms control', *SIPRI Yearbook 2002* (note 27), pp. 506–11.

[156] Kile (note 155), pp. 490–500.

[157] See 'Attachment A: Unclassified report to Congress on the acquisition of technology relating to weapons of mass destruction and advanced conventional munitions', 1 Jan./30 June 2003, URL <http://www.cia.gov/cia/reports/721_reports/jan_jun2003.htm#5>.

[158] See, e.g., Graham, B., 'Missile defense testing may be inadequate', *Washington Post*, 22 Jan. 2004, p. A4; and Richter, P., 'Missile defense system doubts', *Los Angeles Times*, 22 Jan. 2004, p. A8.

Council (GCC) have examined options for a common ballistic missile defence capability.[159]

It is an interesting paradox that the USA's current commitment to international cooperation on missile defence technologies involves a considerable sharing of sophisticated systems while the MTCR seeks to restrict the same or similar technologies down to every conceivable unmanned vehicle capable of delivering a small weapon payload, even to a 100-km range. Missile defence will involve the development and testing of highly manoeuvrable, fast missiles with sophisticated guidance and control systems. These technologies are bound to diffuse to many technically capable countries once missile defence becomes an accepted component of legitimate international activity. The need for international cooperation is also relevant because the sphere of activity goes beyond the limits of any one sovereign country.

From a technical point of view missile defence technologies are grouped according to their application range, that is, short-, medium- or long-range. For terminal defence, the US PAC (Patriot Advanced Capability)-3 system is only usable on short-range missiles because long-range systems would travel much faster in their terminal phase.[160] Extended Range Interceptor missiles and the Tactical High Energy Laser may provide a more viable solution for point defence, when ready. Future plans may include the multinational Medium Extended Air Defense System (MEADS) for defence against 300- to 1000-km range missiles as a cooperative programme between the USA and some other NATO countries.[161] For defence against missiles with a range of 1000–3500 km, the US Army's Theater High-Altitude Area Defense (THAAD), and the Navy's Standard SM-3 missiles are mid-course interceptors that have had varying degrees of success.[162]

---

[159] See, e.g., Kahwaji, R., 'Gulf Cooperation Council threat perceptions and deterrence objectives', *Comparative Strategy*, vol. 22, issue 5 (Dec. 2003), pp. 518–19. The member states of the GCC are Bahrain, Kuwait, Oman, Qatar, Saudi Arabia and the United Arab Emirates.

[160] For more on the PAC-3 see Missile Defense Agency (MDA), 'PATRIOT Advanced Capability-3 (PAC-3)', fact sheet, 30 Jan. 2004, available at URL <http://www.acq.osd.mil/mda/mdalink/pdf/pac3.pdf>.

[161] For more on MEADS see Missile Defense Agency (MDA), 'Medium Extended Air Defense System (MEADS)', fact sheet, 30 Jan. 2004, available at URL <http://www.acq.osd.mil/mda/mdalink/pdf/meads.pdf>.

[162] For more on THAAD see Missile Defense Agency (MDA), 'Terminal High Altitude Air Defense (THAAD)', fact sheet, 1 Mar. 2004, available at URL <http://www.acq.osd.mil/mda/mdalink/pdf/thaad.pdf>.

Since all the currently available options are essentially intended to intercept a single threat at a time, the possible use of countermeasures such as decoys or sub-munitions remains a major problem.

For boost-phase defence, the US airborne laser (ABL) programme envisages high power lasers (HPL) with a 400-km range on board a modified Boeing 747 aircraft capable of engaging ballistic missiles and destroying them with intense heat generated by laser radiation. While yet to be tested as a complete system, DEW technology offers the unique advantage of speed-of-light engagement over long distances, particularly outside the dense atmosphere. Although several problems relating to adverse weather and atmospheric distortions are yet to be resolved convincingly, this project will most certainly lead to the induction of a totally revolutionary technology, DEW, for missile defence and other applications.[163]

Long-range defence could also make use of the powerful interceptors that are at present undergoing development trials as part of the US BMD programme. For mid-course engagement outside the atmosphere, the US Administration may also revive the kinetic energy interceptor (Brilliant Pebbles)[164] programme and the space-based laser programme. The priority attached to these and other projects by the Bush Administration can be measured by its budget request of $7.67 billion for ballistic missile defence for fiscal year 2004.[165]

Despite its many unresolved issues, the robust missile defence effort by the USA can certainly be taken as a measure of the US Administration's perception of the lack of success of the MTCR. Indeed, the MTCR can only prevent missile proliferation where there is a political consensus, and perhaps slow down some indigenous programmes in some states. It is possible that this was its primary intention and that it was not really geared to comprehensively address all missile proliferation. However, the history of the MTCR certainly

---

[163] See the Internet site of the Federation of American Scientists, 'Airborne laser', 13 May 2003, at URL <http://www.fas.org/spp/starwars/program/abl.htm>; and 'Integrated testing of first airborne ray gun completed', SpaceDaily, 22 Apr. 2004, URL <http://www.spacedaily.com/news/laser-04f.html>.

[164] The Brilliant Pebbles system was part of the missile defence architecture of the Administration of President George H. W. Bush (1989–93), known as Global Protection Against Limited Strikes (GPALS). It was to consist of 500–1000 hit-to-kill interceptors. Each interceptor would be housed in an orbiting satellite which would provide communications with ground stations. See Pike, J., 'The military uses of outer space', *SIPRI Yearbook 2002* (note 27), pp. 653–54.

[165] Kile (note 115), p. 648. The US fiscal year 2004 runs from 1 Oct. 2003 to 30 Sep. 2004.

seems to support the argument that a control regime that is narrowly conceived, and implemented in an overtly discriminatory way, is bound to fail in the long run.

There is an argument that aggressive missile non-proliferation efforts can assist missile defence programmes by limiting the technological sophistication available to the adversary. However, adversaries are now difficult to define and the USA is apparently in favour of missile defence cooperation with Russia and has assured China that BMD is not directed against it. If BMD is to be understood as primarily oriented towards threats from rogue states, its elaborate character is hard to justify since the technological sophistication of these threats would not be challenging.

A more widespread view in the world is that BMD will spur missile technology competition among strong sovereign nations on a new scale, because it brings in new concerns related to the security of outer space as well as unknown dangers to individual nations' space assets, which are increasingly vital in the new network-centric world. From this point of view, international cooperation on missile defence technology is highly desirable if the target can be confined to rogue states and non-state players. How such cooperation can be reconciled with the objectives of the MTCR is an issue that will need serious techno-political negotiation with some of the former target nations and may well decide the future of missile non-proliferation efforts.

V. The efficacy of multilateral export control regimes

The present export control regimes have no universal treaties, nor can they claim to be non-discriminatory in terms of meeting the legitimate needs of all sovereign states. At present, there are four main multilateral export control regimes that essentially represent the international arrangements that have evolved on the basis of shared perceptions of threats and security over the past five decades. These are the Australia Group, the Missile Technology Control Regime, the Nuclear Suppliers Group and the Wassenaar Arrangement. Unlike formal treaties, they started off as informal supplier arrangements—restricted clubs. Over time, the participants became more organized and expanded to include more participant states, which may or may not have been suppliers with the same perspectives or interests as the founders. These regimes have served their purpose fairly well. They have played an important role in containing direct acquisition-based proliferation and in slowing the progress of indigenous technology development in the countries targeted.

The rules developed in these international control regimes currently apply to five main categories of militarily applicable items and services—WMD, major weapon systems, light weapons, dual-use technologies, and products and services such as training. Each category has its specific peculiarities requiring different control methodologies. There is general consensus and a fair amount of clarity about controlling WMD technologies. As regards conventional weapons, the dual-use items associated with them and other items of high technology that may have military applications, there is only tenuous agreement about how to apply current controls, that is, which countries to deny and which to supply.

The Wassenaar Arrangement

With the end of the cold war, export controls were left with no clear targets and, as discussed above, in 1992 the 17 participating states in COCOM decided to adapt and supplement their arrangement. In 1993 the participating states took a further step and decided to abolish COCOM and replace it with a new arrangement with a focus on technology control based on mutual coordination. Five rounds of meetings were held in Wassenaar, the Netherlands, to define the scope and structure of the new multilateral export control regime that was to control the trade in conventional weapons and dual-use, high-tech items. Some might conclude that the WA was evolved not out of any compelling security concerns but rather out of the need to retain techno-economic superiority and a desire to be in control of high-tech matters of military importance in the post-cold war power system. While this may not have been the genuine rationale, it certainly represents a perception among the target countries, such as India, that are denied legitimate access to dual-use technology.

Although conceived as a new and open system, responding *inter alia* to lessons from Iraq and the 1991 Gulf War, participation in the WA was as important as its scope and purpose. After much debate, interlaced with disagreements and deadlocks, the WA was established on 12 July 1996 as the new forum 'to contribute to regional and international security and stability by promoting transparency and greater responsibility in transfers of conventional arms and dual-use goods and technologies, thus preventing destabilizing accumulations'.[166]

The WA, conceived, structured and established after the cold war, is interesting to study because it gives insight into West-dominated thinking on how technology controls should be managed in the changed strategic environment of the future. The negotiations leading to the formation of the WA demonstrated a recognition that the discriminatory practices of the cold war were no longer relevant or workable and that future export control regimes would have to be far more sensitive to economic competitiveness, international trade and commerce. The WA grew to include 33 participants (see table 3.1) with some of the old target countries becoming partners and a new definition of the target group emerging. Countries sought to join the WA in order to remain influential in international matters and to gain better access to high-tech commerce.

The option of identifying specific countries for closer scrutiny was discussed and advocated by some states (principally the USA). There was no general agreement, however, on the identity of states that should be targets for export controls and none is named in the founding documents (the Initial Elements).[167] As stated above, of the 73 sensitive destinations that could be identified in the lists and guidelines of four founding participants in the WA (Germany, Japan, the UK and the USA), only 28 were found to be common to all. Thus, although the WA retained such features of COCOM as the classification of controlled items under the munitions list and the dual-use list, it finally evolved as a much more open system with export decisions left to national discretion and no veto powers. Information sharing and transparency are the pillars of the WA, with participating countries agreeing to inform each other within 60 days about approvals of licences to export items that resemble those where a licence was denied by another participant in the preceding three years. This post-shipment notification stops short of a no-undercut policy.

---

[166] The Wassenaar Arrangement on Export Controls for Dual-Use Goods and Technologies, 'Initial elements', quoted in Lipson, M., 'The reincarnation of COCOM: explaining post-cold war export controls', *Nonproliferation Review*, vol. 6, no. 2 (winter 1999), pp. 38–39. The full text of the Initial Elements is available at URL <http://projects.sipri.se/expcon/wass_elements.htm>.

[167] Boese, W., 'Divisions still impede Wassenaar export control regime at plenary', *Arms Control Today*, vol. 27, no. 8 (Nov./Dec. 1997), p. 27, available at URL <http://www.armscontrol.org/act/1997_11-12/wassnov.asp>.

The difficult negotiations on setting up the WA are testimony to the reality that export control cooperation can only be expected as a collective response to common threats or as the result of coercion by a dominant state. International cooperation on controls for conventional arms and dual-use technologies is consequently far less forthcoming in present-day conditions than for WMD technologies. The arms export trade is a multi-billion dollar business that sustains the national economies of many supplier states, and commerce in high technology is becoming increasingly competitive.

Nevertheless, and while the WA continues to operate in an informal manner, it has now developed certain common perceptions regarding countries of concern. Although no state or region is named as a target of the WA, over time the discussions among participating states have led to understandings on restraining arms transfers to particular regions (such as Central Africa and West Africa) and particular countries (such as Afghanistan and Sudan). Since 1999 the chair of the WA General Working Group has coordinated a more structured information exchange on the basis of global and regional views (contributed by participating states). These describe and analyse the pattern of arms acquisitions in particular countries and regions of particular concern to the state that submitted the document to the group.[168] From the perspective of progressive developing economies such as China or India, this evolution of the WA may look very much like the strengthening of an informal supply cartel, which denies technologies based on political or economic priorities linked to techno-economic security and to the desire to maintain a competitive edge.

In any event, and however the motives behind the WA may be interpreted, the realities of the trade in conventional arms and dual-use technologies on the ground are quite different. Neither the UN Register of Conventional Arms nor the WA has been able to prevent several controversial arms transfers by powerful nations. Seen from the demand side, export controls thus seem to lack a demonstrated security rationale. They continue to appear selective, discriminatory and highly political and to show inadequate sensitivity to the regional implications or consequences.

---

[168] Anthony, I., 'Multilateral weapon and technology export controls', *SIPRI Yearbook 2000: Armaments, Disarmament and International Security* (Oxford University Press: Oxford, 2000), pp. 672–80.

The WA is meant to prevent the regionally destabilizing acquisition of conventional weapons, but there can be any number of arguments for or against any regional weapon acquisition depending on the political considerations prevailing at the time. To be coherent, such an objective must recognize that it is governments or dictators, not weapons, that wage wars. The 'destabilizing factor' is the product of intentions and capabilities. The real focus should therefore be more on the recipient states' record of responsible behaviour, as well as their political configuration and goals. Unfortunately, the WA continues to focus primarily on lists of items and on the perceptions of a limited range of participating states, rather than attempting an unbiased analysis of the recipient's credentials.

From the supply-side viewpoint, arms export controls and technology controls certainly have a role to play in the security calculations of modern societies. This is not a matter of dispute for the demand side either. However, for the demand side, excessive, biased or unbalanced controls not only obstruct the path to development and progress but also are detrimental to regional security perceptions. The supply side has traditionally failed to have adequate sensitivity for this viewpoint. During the cold war this was understandable because adversarial policies and export controls enjoyed a fair degree of success based on the clear identification of 'us versus them' and because of the clear perception of the looming threat of a major war of the worst kind. In the 21st century, questions arise as to whether similar export controls are still relevant in such dramatically changed circumstances and, if so, how much control is optimal for the future. What should be the priorities of a regime such as the WA, now that it has former enemies as participants and possible future partners as today's targets?

There is a universal consensus on the need for technology controls for curbing terrorism and violence. However, it is possible to argue that, having invested so much in the creation of structures and procedures for export controls, it is difficult for regime participants to give all this up as irrelevant. There is therefore a tendency to maintain the organizations for eventual future use, perhaps more for economic benefit than for security reasons (especially in the areas of dual-use technology controls). The latest efforts to give export controls a clearer non-proliferation profile could be seen as an attempt to retain the high moral ground as justification for continuing control activities.

An alternative argument is that regional instabilities can seriously affect the peace and well-being of the major developed nations and that export controls can help to reduce such instability by maintaining control over regional situations.

How effective are these multilateral export control regimes anyway? Although the US-led export controls against the Soviet bloc countries proved fairly effective in the final analysis, the lesson of the 1991 Gulf War was an eye-opener as regards the weaknesses of the system vis-à-vis other regions. Much of Iraq's WMD capability arose initially from purchases of components and sub-systems by Iraq's agents, who successfully exploited the loopholes in the export control systems. It must also be remembered that Iraq had previously been an ally of the USA and was familiar with ways to conduct defence-related business with the USA and its allies. Recent findings regarding WMD-related acquisitions by Iran, Iraq, and North Korea indicate that these activities were supported significantly by 'export decisions, deliberate or inadvertent, made by supplier states such as Russia, China, Germany, France and Pakistan'.[169]

The coordination of technology controls did go a long way towards harmonizing the export controls of major Western supplier states, and the present control regimes can certainly take credit for containing technology proliferation to avert an even worse situation. However, the regimes also have many drawbacks and in today's changed circumstances they are grossly inadequate to meet the new requirements. First, the regimes operate informally on the basis of consensus, where any one participant can block the process. Second, it is left to individual participant states to implement the collective decisions taken, within a certain time frame. Third, former Soviet states such as Ukraine and others now participate—making a rather disparate group with different priorities on economic and strategic issues. Participating states suffering economic hardship often find it difficult to deny themselves the commercial benefits of lucrative arms deals that can somehow be made to fit within their own interpretations of export controls. Fourth, many new participants, and even European allies of the USA, often find US regulations far too strict. Some perceive the controls as a means of maintaining the significant technological edge that the USA enjoys. Some of the new participants lack either the

---

[169] Beck and Gahlaut (note 134).

necessary infrastructure or the political will to fully implement US standards for export controls.[170]

Export controls have therefore succeeded best where there was overarching agreement that their joint security benefits outweighed individual compliance costs. However, when the short-term economic incentive to sell becomes more important than long-term security benefits accruing to an individual state, export controls become restrictive and there are economic incentives to bypass them. Verification and monitoring of compliance are subjects in themselves with as many variations as possible for a large variety of technologies and products as well as for a large spectrum of supplier priorities. There is no one-size-fits-all solution and the potential for ambiguity is fairly high. The cost of non-compliance is also not clearly enunciated in any of the existing control regimes, and nations powerful enough to fend off peer pressure can afford to give priority to protecting their economic interests if the pay-off is large enough.

There is also the point of view of the potential recipient state, which is never a party to the multilateral regime. The receiving state seeks advanced technology to enhance its security and economic development and sees its actions as no less legitimate than the actions of other states during their own period of development and modernization. Hence, in the current context, unless there is clear identification of enemy status, directing export controls towards another state just because it is not a participant in a supply cartel is untenable and should be internationally questionable. Security perceptions and economic competition continue to be legitimate fields and export controls will need to revisit the target country definition in the future.

Another problem area is end-use certification. The responsibility for this lies with the supplier entity (which is usually a private company, and not a government agency) and the action lies at the recipient end where a number of ways exist to violate the spirit of controls. In dual-use or multiple-use generic-technology areas, the applicability of such certification becomes too complex and uncertain to comply with or even to monitor.

It is important to emphasize that the process of export controls is also dependent on various national regulations and their varying

---

[170] Beck and Gahlaut (note 134). For the example of Ukraine see Diamond, H., 'U.S., Ukraine sign nuclear accord, agree on MTCR accession', *Arms Control Today*, vol. 28, no. 2 (Mar. 1998), available at URL <http://www.armscontrol.org/act/1998_03/ukrmtcr.asp>.

interpretations by agencies such as trade, industry or commerce departments and customs authorities, which do not share uniform backgrounds or priorities across nations. In fact, even in the US system differences in perception between the State Department and the Department of Commerce are not unknown.

The idea of strict export controls across a wide spectrum of dual-use products and services is seriously at odds with the reality of the times. It implies essentially that the 'technology haves' will want to exercise control to their advantage and that 'technology aspirants' will see controls as unfair practices and restrictive to their legitimate right to progress and defend. There can never be an ideal solution that will satisfy all. However, an effort is warranted to fine-tune control regimes to the changing realities of the times and focus on real global-level threats to allow dangerous capabilities as well as intentions to be contained. The equally important issue of intent thus assumes a more significant dimension and some tempering of intentions must form an important objective for future technology control regimes. The recent tendency of export control regimes to adopt end-use controls is itself a recognition that assessments of technology need to take into account not only the performance parameters of items but also the intentions of those who will have access to them.

It is in this context that reducing tension and creating wider cooperation internationally take on new meaning for creating a wider and deeper international consensus against common dangers. This consensus can in turn be the basis for universal participation in the efforts to control technology flows—something that is not possible in practice for existing export control cooperation arrangements. The international will and the international voice must become strong enough to act as a deterrent to technology misuse and any form of rogue adventurism.

The way forward might lie in an effort to review and redefine the priorities of future technology controls to make them more effective by a narrower focus on fewer areas that truly pose a threat to international peace and stability. The right prescription may be 'taller fences around fewer technologies' to enable an effective system of controls to operate in parallel with unhindered trade in benign technology for development and progress. This approach, which is consistent with the general trend in export control thinking in recent years, would have many advantages. It would help all nations to focus on real threats to modern society irrespective of individual priorities and economic competition. This would foster much wider and more spontaneous cooperation among nations on the requirements of technology controls and help to evolve international standards regarding acceptable and unacceptable behaviour. It might also help define universal taboos.

The levelling effects of technology in the 21st century will make perceptions and inclinations much more important issues for the interdependent world. Opening the debate on controls on technology to a wider set of actors may reduce the tendency to view export controls as unfair practices engaging only a few powerful nations. The risk that a closed discussion will add to international tensions and mutual mistrust in the long run is already leading to greater outreach efforts by states that participate in export control cooperation arrangements.

A failure to engage in more open debate entails an unacceptable risk that separatist elements and fundamentalist organizations will find political cover for their efforts and may even combine forces—a development certain to breed tension, discord and violence. If such tendencies grow out of hand, the situation will only demand more controls and more punitive actions, fanning further polarization in a world already under stress because of limited resources, and driving a downward spiral that could lead to anarchy. If this is allowed to happen it will be a great missed opportunity for humanity to use the tremendous potential of technology towards the realization of democratic freedoms and peace through a matched distribution of power in the world.[171]

Advanced military capabilities in the possession of responsible democratic states actually act as a deterrent to war, as has been proved in the industrially advanced nations. These nations continue to spend significantly larger amounts on the modernization of their defence forces compared to even the highest spenders in the developing regions. Even with the dramatically diminished threat of conventional war, they continue to invest heavily in advanced conventional military capabilities. In the context of international stability and moving towards lasting peace, the question will be: is the continued acquisition of more and more advanced military technology by any one state

---

[171] See, e.g., Feenberg, A., 'Democratic rationalization: technology, power and freedom', eds R. C. Scharff and V. Dusek, *Philosophy of Technology: The Technological Condition: An Anthology* (Blackwell: London, 2003), pp. 652–65.

or group of states conducive to a positive international security environment? If the answer is a qualified no, then the international arms control institutions must address this question more seriously rather than just by expanding and sharpening the old-style export control mechanisms. The changed circumstances of the 21st century, despite the apparent increase in regional tensions and overall violence, have created a unique opportunity to address arms control in its true sense, rather than mix it with export controls, and to seek new avenues for more lasting international stability and peace.

4. Technology diffusion in the 21st century

I. Technological interplay

During the cold war years, Soviet technology was perceived as broadly comparable to that of the USA in many areas and was seen as the foremost security threat. Coordinated strategic steps were therefore initiated to maintain the US lead where it existed and to overtake the Soviet Union in key technology areas such as advances in nuclear weapons, missile technology, space technology, surveillance technology and stealth technology, to name a few. The NATO alliance was established and the US arms industry was given the funds needed to undertake continuous upgrades to conventional weapon technologies. At the same time, COCOM was set up to deny technological benefits to the adversary and a substantial network of export control mechanisms was put in place to prevent the newer dual-use technologies from reaching countries other than close allies. In the USA, DARPA was constituted to push at the cutting edge of technology for the benefit of US strategic capabilities. The focus of US security policy during the 1970s and 1980s was on rapid technological development and economic progress at home, while erecting every possible barrier to prevent technological know-how from leaking to the outside.

Looking back, other nations on the technology-acquisition curve have a lot to learn from the US model because it imposed a heavy economic burden on an adversary that contributed to the defeat of the latter. This victory did not involve fighting a war—a significant achievement given the history of mankind. Techno-economic competition therefore stands proven as one of the most effective tools for defeating a political or military adversary.

The US victory over the Soviet Union was, however, not without problems. Many new technology-related security concerns emerged from the disintegration of the Soviet Union. Of major concern was the safety of the large stockpile of nuclear weapons and fissile materials. The stockpiles of the two superpowers had already exceeded the limits of economic viability and safety management, and several initiatives for mutual disarmament were in progress even before 1990. This provided some background to and familiarity with the issues of rapid disarmament and related safety issues. A good account of this effort exists in the open literature.[172]

Technology diffusion may be defined as the natural spread of technology through every type of technology interaction, whether acquisition, development, transfer, co-production or even intellectual exchange. A major problem after the break-up of the Soviet Union was the vast bank of knowledge invested in Soviet scientists, who lost their privileged access to state resources at a time when political changes made it easier for them to establish connections with foreign potential buyers of their expertise. The risk that this knowledge would spread according to the logic of market forces led the USA to initiate a major project to absorb and rehabilitate these scientists. Other countries, such as China and Israel, also used the opportunity to their advantage. The importance of technological knowledge for security was again vividly demonstrated. The ubiquitous nature of technological knowledge is largely responsible for technology diffusion in 'intangible' ways that are not easily obvious to monitoring and control agencies. This, in turn, also contributes invisibly to the process of technology diffusion.[173] It is interesting to note that, while the West was occupied with the management of dangers related to the catastrophic failure of the Soviet system, China was quick to learn from immediate history and used the so-called peace dividend to maximum advantage. China reviewed its military modernization efforts, under way since the 1980s, and chose to concentrate fully on reorienting for techno-industrial superiority to match the best in the world and on building economic competitiveness to overtake its powerful neighbour—Japan. Throughout the 1990s, China continued on its path of rapid economic growth and consolidated its military and commercial technology base. The rise of China as a potential world power is largely due to the impressive growth in its manufacturing base and the globally competitive edge that China has gained in the process.

---

[172] See Anthony (note 49).
[173] For a discussion of intangible transfers see Anthony, I., 'Multilateral weapon and technology export controls', *SIPRI Yearbook 2001* (note 122), pp. 631–35.

II. The impact of new technologies

Increased globalization and rapid advances in technology have weakened national borders and enhanced technology diffusion. The IT revolution and the spread of individuals' skill-oriented knowledge make export controls almost impractical in some areas. Applying export controls to a large band of technologies and to all countries requires a significant infrastructure to help make licensing assessments if implementation is to be effective. The associated costs, combined with the opportunity costs from lost export earnings, could make export controls too expensive to justify for many nations.

As discussed in chapter 1, technology diffusion and increased globalization have made international transactions far more interdependent and market-driven than before and this trend is bound to accelerate as economic competition becomes sharper. Even clear-cut arms export controls have faced problems because of the pressures of the arms-export industry. When it comes to controlling dual-use technologies, definitions of what can be exported safely, and to whom, become even more complex. For example, Germany and Sweden are believed to have sold industrial electron beam machines to the Semiconductor Manufacturing International Corporation (SMIC), a Chinese manufacturer of computer chips. The USA, however, is known to have banned such exports to China.[174] It is open to question whether this is a case of undercutting the USA or of disagreement over how to interpret agreed export control guidelines. In an age of rapidly advancing technology, how far is it practical to control industrial machine-tools that can enhance technological capabilities in the long run? Given the rate of obsolescence in high-tech fields, what is new today may be commonplace next year and therefore too difficult to control because of the large number of possible suppliers.

The technology denial regimes of the past five decades have spurred indigenous technology growth in many progressive countries. One of the major problems for export control regimes is the realization that a number of countries outside the regime, such as China, India and Israel, have become potential technology suppliers themselves. In addition to being important techno-economic players of the future, these countries are also potential markets for sales of high technology.

---

[174] Read, R., 'US trade, security interests clash over tech exports to China', *Oregon Live*, 3 Feb. 2003, cited in Beck and Gahlaut (note 134).

Countries such as North Korea and Pakistan represent a different group of nations that do not have the techno-economic strength to be future international players but are capable enough in certain sensitive defence technologies to upset many international non-proliferation objectives. In another 10 years, several other countries may be able to export important technological goods and services. These countries do not have the same technology absorption capabilities and will therefore react differently to non-proliferation priorities.

Key technologies: information technology, biotechnology and energy

The 20th century began with the birth of aerospace technology, which subsequently revolutionized war doctrines and security perceptions during two world wars. In the past five decades phenomenal achievements and technological advances have been made in space technologies, which have added the new dimension of outer space to security and threat perceptions. At the same time, key enabling technologies, such as semiconductors, integrated circuits, computers, lasers and photonics, have greatly enhanced overall technological capabilities and system performances. Apart from the WMD technologies that have evolved fairly quickly over the past 50 years, it is these technologies that shape the world we live in and influence all aspects of life—including our sense of security and well-being. Given the fast pace of technological change, it would be perfectly natural to expect significantly new technologies to find their way into everyday life in the 21st century, including some technologies which will affect security and threat perceptions. IT, biotechnology, energy technologies and space technologies are obvious major contenders that deserve closer examination and appreciation.

The IT revolution and the growth of fast, compact computers have changed lifestyles in most modern, progressive nations. However, this new level of dependence on digital electronics and IT in every walk of life, including defence and security matters, has brought about a new vulnerability to, and a consequent new threat perception from, the risk of cyber-warfare at many different levels. Unlike conventional military hardware that causes destruction and death, cyber-warfare techniques use intangible software tools that can cripple military capabilities and international commercial trade. In a sense they are full-spectrum techno-economic tools for use in both defensive and offensive strategies. The nature of this technology makes individual brainpower more relevant than techno-industrial infrastructure, thus threatening to compromise the huge technological advantage that the Western industrialized nations have established after years of effort. There are now new types of threat to an information-based society, where information security becomes perhaps as important an issue as defence against WMD attack. Technology has thus changed the nature of warfare from obvious and visible large-scale military action and violence to subtle, invisible yet decisive capabilities for crippling the enemy's information environment in a warlike situation, thus denying it command, control and communications (C^3) advantages.

Today's defence strategies are already heavily influenced by such new vulnerabilities, and the use of information is now a tool for achieving military and economic objectives. The ubiquitous nature of information systems has also removed the clear distinction between covert and overt actions, because there is no clear, common, international agreement or even understanding of the acceptable and legitimate limits of using IT to protect national security interests. Paradoxically, it is the advances in sensor technologies and enhanced IT capabilities that are also responsible for enabling the technological edge necessary to counter WMD threats. IT can aid comprehensive monitoring and verification techniques for compliance verification as well as for early detection of proliferation activities, thereby complementing national technical means (NTM) for verification and monitoring. This will be valuable for the verifiable reduction or elimination of WMD arsenals and thus enhance confidence among the countries participating in cooperative disarmament. It is to be hoped that this will lead the world towards meaningful universal disarmament. Should such technology be controlled and, if so, how could this be achieved?

Since the late 1970s there have been significant advances in the understanding of the science of life and this has prompted a surge of investment in biotechnology research that will not only lead to medical advances but also make it possible to introduce genetic changes to food crops to bring about higher yields and better resistance to dis-

TECHNOLOGY DIFFUSION 107

ease.[175] The combination of IT and biotechnology has allowed substantial genome sequencing and analysis, creating a wealth of information in the areas of health care, food production and agriculture. While so many benefits are brought forth by the advances in the life sciences, the greater understanding of the processes that underpin life raise several moral and ethical concerns about biological warfare.[176] A combination of biotechnology and nanotechnology may help to realize unprecedented miniaturization and new capabilities that could be used for both constructive and destructive purposes.[177] Like most advanced technologies, biotechnology is a double-edged sword that can either heal or hurt, depending on how the technology is managed. Either way, the implications for the future are enormous, and careful, universal and coordinated management will be crucial.

Another area of potential impact is energy technology. Future possibilities could include an alternative cheap and abundant source of energy that would not only revolutionize everyday life but also shift the strategic balance of oil-dependent economies. Research on controlled thermonuclear fusion could provide unlimited energy from seawater.[178] Similarly, future research on hydrogen fuel may revolutionize the automobile industry and propulsion technologies.[179] This cutting-edge research work would also facilitate the realization of practical, affordable DEW that could totally revolutionize conventional warfare, introduce new dimensions to threat perceptions and alter security calculations.[180] These examples are only a few sample possibilities. Several such technologies that could create opportunities for quantum leaps in techno-military capabilities are bound to affect the future. Most of these potential future technologies will be in the

[175] For a useful introduction to the field of biotechnology and its consequences see Moses, V. and Cape, R. E. (eds), *Biotechnology: The Science and the Business* (Harwood Academic Publishers: London, 1991).

[176] See, e.g., Dando, M., *Biological Warfare in the 21st Century: Biotechnology and the Proliferation of Biological Weapons* (Brassey's: Dulles, Va., 1994).

[177] See, e.g., Moodie, M., Chemical and Biological Arms Control Institute (CBACI), *Reducing the Biological Threat: New Thinking, New Approaches*, Special Report no. 5 (CBACI: Washington, DC, Jan. 2003).

[178] While nuclear fusion as a power source may be possible in the future, no programme has yet advanced beyond the research stage and practical power generation will not be achieved for a number of decades.

[179] Hydrogen fuel has a number of advantages, including pollution-free driving. However, the production of the hydrogen itself will consume large quantities of energy. The hydrogen economy will therefore be a revolution in how energy is delivered to the user, not in the sources of energy used around the world.

[180] See note 163.

dual-use domain and some of them are more than likely to be within reach of many progressive nations.

Diffusion of outer space technology

One such application of new technology with an immense potential impact on the future will be the use of outer space. Since the Russian launch of Sputnik in 1957, space exploration has captured the imagination of many nations and past decades have witnessed impressive advances in the capacity to use outer space for peaceful civilian purposes as well as for military reconnaissance, navigation and communication. The 1967 Outer Space Treaty[181] envisaged the exploration and use of outer space for the benefit and in the interest of all countries and recognized outer space as beyond national boundaries, subject to internationally agreed rules of conduct. However, technological advances have opened up the potential for significant military uses of outer space. The present plans for BMD have several technology components that could be used against satellites, particularly those in low earth orbit (LEO).[182] Interceptor missile tests could increase the level of space debris, thereby creating hazardous conditions.[183] The LEO satellites and new-generation DEW can also have anti-satellite (ASAT) capabilities. There appears to be inadequate debate or international dialogue on the security implications of the imminent introduction of weapons into space. If current developments either set off a new race for military capabilities in space or provide highly selective benefits to one country, in violation of the spirit of the Outer Space Treaty, the overall net effect on security may be negative.

The low earth orbit is a shared venue for scientific exploration, commercial communications, navigation and remote sensing. The present US missile defence plans will use the same space for military engagement with enemy missiles using kinetic energy interceptors. Military-specific early-warning satellites, missile-defence sensors and tracking radars will also be deployed under the US BMD programme.[184] All this will be for the selective benefit of one country

---

[181] See note 11.
[182] See Pike (note 164).
[183] Pike, J., 'The paradox of space weapons', *SIPRI Yearbook 2003* (note 6), pp. 433–38.
[184] The Report of the US Commission to Assess United States National Security Space Management and Organization, Washington, DC, 11 Jan. 2001, URL <http://www.defense

against another, in violation of the spirit of the Outer Space Treaty. Technological capabilities continue to be developed under generously funded projects without any international clarity about the future of weapons in space. This could lead to a situation in which space technology becomes a victim of unwarranted secrecy and unilateral controls in the absence of serious debate either at the UN or within the country responsible. An international debate on the future use of space for military purposes should not only address the issue from the perspective of an arms control and non-proliferation framework, but also bring about the clarity needed to reach international consensus regarding acceptable norms for the use of space for activities such as missile defence or even satellite defence in future.

As noted above, the USA withdrew from the ABM Treaty as of 13 June 2002, clearing the way for it to develop and deploy nationwide defences against long-range ballistic missiles. Limited defences against short- and medium-range missiles were already allowed under the ABM Treaty, but the USA's withdrawal gave it the freedom to develop, test and deploy any or all forms of defence system it may deem fit in the interests of its national security. On 17 December 2002, President Bush announced plans for an initial BMD deployment.[185] They include deployment of 10 ground-based interceptor (GBI) missiles by 2004, of which six would be located at Fort Greely in Alaska and four at Vandenberg Air Force Base in California. Another 10 GBI missiles are planned for 2005, with 20 more sea-based interceptors on three ships and an undisclosed number of PAC-3 short-range interceptors. The long- and medium-range interceptors will utilize exo-atmospheric kill vehicles whereas the Patriot interceptions will be endo-atmospheric.[186] In a document released on 20 May 2003, the Bush Administration described the 2004 deployment as a 'starting point in an evolutionary approach to missile defence', indicating that there may be no fixed or final missile

---

link.mil/pubs/space20010111.html>, calls for the deployment of space weapons for both missile defence and satellite defence.

[185] 'President announces progress in missile defense capabilities', The White House, Office of the Press Secretary, Washington, DC, 17 Dec. 2002, URL <http://www.whitehouse.gov/news/releases/2002/12/20021217.html>.

[186] For more detail see Boese, W., 'Missile defence post-ABM Treaty: no system, no arms race', *Arms Control Today*, vol. 33, no. 5 (June 2003), pp. 20–24.

defence architecture and that the USA is keeping options open for the future.[187]

Ballistic missiles are categorized according to their range. Short range is usually defined as up to 1000 km, medium range as 1000–3000 km and intermediate range as 3000–5500 km. Missiles with a range of 5500 km or more are defined as intercontinental ballistic missiles (ICBMs). The longer the range, the higher the trajectory and the higher the speed. Long-range missile warheads are capable of travelling at speeds of over 7 km per second. Missile defence envisages the interception of enemy missiles in either their boost phase, mid-course phase or terminal phase. Long-range ICBMs may take three to five minutes for their boost phase, up to 20 minutes for mid-course ballistic unpowered flight, depending on range, and a final few minutes of terminal flight after re-entering the earth's atmosphere—hitting targets at speeds of over 3200 km per hour.[188] Missile defence systems for the various stages will thus have to rely on appropriate technologies to meet the full spectrum of anti-missile engagement. Levels of technological sophistication and system performance will obviously be demanding if missile defence is to have a chance of succeeding against advanced attack missiles in the short time available for reaction.

Current US BMD plans include boost-phase intercept using an airborne laser. This will be the first of its kind, using a speed-of-light energy beam weapon at high altitude to destroy enemy ballistic missiles in their initial boost phase from a range of over 400 km. The ABL will be a modified Boeing 747 aircraft with a megawatt-level chemical oxygen iodine laser (COIL) that can cause fatal structural damage to enemy missiles using a short burst of laser energy by direct line-of-sight, instant targeting through the thin layers of the atmosphere, at an altitude of over 10 km, as the missile breaks through the cloud top. The high power laser under contract with TRW Instruments has made impressive progress and the first ABL platform was flight-tested in July 2002. Full operational tests are planned for 2005 at the

---

[187] 'National policy on ballistic missile defense fact sheet', The White House, Office of the Press Secretary, Washington, DC, 20 May 2003, URL <http://www.whitehouse.gov/news/releases/2003/05/20030520-14.html>.

[188] For a basic description of ballistic missile technology see the Internet site of the Federation of American Scientists (FAS), FAS Special Weapons Primer, 'Ballistic missiles' at URL <http://www.fas.org/nuke/intro/missile/index.html>. See also Karp, A., SIPRI, *Ballistic Missile Proliferation: The Politics and Technics* (Oxford University Press: Oxford, 1996), pp. 99–146.

earliest, but problems with the weight of the modular laser system and atmospheric distortion of the high-power laser beam at ranges of over 400 km are yet to be resolved.[189] However, given the technology indicators and the funds already committed, eventual deployment of an ABL using DEW appears inevitable.

The impact of this new class of weapon on security perceptions will be significant because it opens up several new vulnerabilities. Apart from the obvious effect of altering the strategic balance of deterrence for several nations, a DEW capability onboard an aircraft above the immediate atmosphere makes all satellites vulnerable. Satellites by design have highly sensitive devices and sensors that can be easily destroyed by powerful laser radiation. The loss of valuable satellite assets would be crippling to most modern nations.[190] Future developments in DEW technology may also allow its use in the tactical theatre of war, where even the most modern conventional weapons might have little chance against speed-of-light energy weapons.[191] This could completely revolutionize the defence and military strategies of most nations.

Other components of the USA's planned BMD system include the space tracking and surveillance system (STSS), space-based infrared systems (SBIRS–High), THAAD, ground-based mid-course defence and the PAC-3 for advanced air defence.[192] Each of these components draws on the latest technology in its respective areas and will create new techno-military capabilities in the future. Obviously, each of these technologies will be attractive to most technologically competitive nations and the accentuated technology race will have its own security reverberations. The exo-atmospheric missile interceptors also raise other important issues, such as additional space debris in LEO, which would increase incidents of accidental damage to LEO satellites. There are already over 9000 space objects of more than 10 cm in size being tracked internationally to protect satellite assets engaged in vital peacetime activities.[193] Weapons in space will not

---

[189] See Federation of American Scientists (note 163).

[190] Pike (note 183).

[191] See Federation of American Scientists (note 163).

[192] The components are described in Missile Defense Agency (MDA), MDA Link, 'Fact sheets', URL <http://www.acq.osd.mil/mda/mdalink/factsheet.html>; and Pike, J., 'The military uses of outer space', *SIPRI Yearbook 2002* (note 27), pp. 647–54.

[193] Moltz, J. C., 'Reining in the space cowboys', *Bulletin of the Atomic Scientists*, vol. 59, no. 1 (Jan./Feb. 2003), pp. 61–66, URL <http://www.thebulletin.org/issues/2003/jf03/jf03moltz.html>.

only increase the chances of space accidents but also require a total review of existing international norms on the peaceful use of outer space.

Advances in technology have also opened up several other possible strategic uses for outer space apart from missile defence. According to recent reports, US plans include reusable unmanned hypersonic cruise vehicles that would fly at over 10 Mach (10 times the speed of sound in air) through outer space to strike targets over 10 000 km away in less than two hours.[194]

Outer space capabilities could also create significant potential for the use of weather as a force multiplier in a few decades. According to a 1996 study by the US Air Force,[195] future weather modification systems could provide unique opportunities for advanced military operations across the full spectrum, enhancing friendly operations and disrupting those of the enemy by tailoring weather patterns for dominance of global communications and surveillance capabilities for counter-space control. As well as advanced weather modelling, reliable weather database generation and global sensor array systems, the technologies involved could also include weather intervention techniques. Some of these technologies already exist and others will

---

[194] The USA is planning a number of military uses for outer space, including high-speed manned and unmanned attack 'aircraft'. See, e.g., Federation of American Scientists (FAS) Military Analysis Network, 'HyperSoar: Hypersonic Global Range Recce/Strike Aircraft', 24 Dec. 1998, at URL <http://www.fas.org/man/dod-101/sys/ac/hypersoar.htm>. For detail on the Common Aero-Vehicle (CAV), described as a manoeuvrable re-entry vehicle with several payload options (including small smart bombs, powered Low Cost Autonomous Attack System munitions, a 'hard and deeply buried target penetrator', a deployable UAV hunter/killer package and an agent defeat payload), see FAS, 'National Security Space Road Map: NSSRM: Conventional Ballistic Missile (CBM) with Common Aero-Vehicle (CAV)', 12 July 1999, at URL <http://www.fas.org/spp/military/program/nssrm/initiatives/cbmcav. htm>. For details of the hypersonic cruise vehicle (HCV) see Schafer Corporation, 'Military spaceplane' at URL <http://www.schafercorp.com/Company/sde/msp.htm>; and FAS Space Policy Project, Military Space Programs, 'Military spaceplane X–40 Space Maneuver Vehicle Integrated Tech Testbed', 14 Jan. 1999, URL <http://www.fas.org/spp/military/program/ launch/msp.htm>. The CAV could be launched using a ballistic missile booster, from the HCV or from a military space 'aircraft'. The Enhanced Common Aero-Vehicle (ECAV) is a longer-range version of the HCV. The military space aircraft, which would actually operate in space, could be used *inter alia* as a launcher for CAVs. However, it should be noted that most of these plans are at an early stage and many programmes could be merged in the future. The first orders for the development of technologies for the CAV, the ECAV and the HCV were placed in 2003. See Vantran, K. L., US Department of Defense, American Forces Press Service, 'FALCON Phase 1 contractors selected', 22 Dec. 2003, URL <http://www.defenselink. mil/news/Dec2003/n12222003_200312221.html>.

[195] House, T. J. *et al.*, 'Weather as a force multiplier: owning the weather in 2025', Paper presented to Air Force 2025, Aug. 1996, URL <http://www.au.af.mil/au/2025/volume3/ chap15/v3c15-1.htm>.

evolve through the pressures of civilian needs for weather prediction for a variety of applications. Space technologies are relevant to the implementation of new and experimental techniques for weather modification. These enhanced technological capabilities will certainly be available for other possible applications, with defensive and offensive objectives, and will therefore add new dimensions to the security perceptions of individual nations in the 21st century.

Technology will continue to be a major player, not only in terms of the sophistication of military systems but also in the area of enabling civilian technologies of wide impact. For instance, new sources of clean, affordable and abundant energy would alter the strategic importance of oil resources; low-cost space-launch technologies could revolutionize the use of space; and the IT revolution could radically transform methods of governance and law enforcement, at both the regional and international levels. All these will affect the distribution of wealth and power and increase the interdependence of nations.

III. Future trends in technology and strategy

The examination of technological interconnections between security and threat perceptions clearly brings out the important role that technology has played in the past and the inevitability of its increasing impact in the future. However, technology, or the military hardware produced by it, is in itself only a means, not an end. Security planning and military strategies as well as the organizational infrastructures to implement the strategies are the key to success. That said, it is technology that provides military planners with a variety of options in line with the range of techno-military capabilities available. The sudden end to cold war calculations provided an opportunity for most nations to review their individual security doctrines and military strategies. The past decade has thus been one of introspection and self-evaluation for many progressive nations, giving them occasion to assess their existing potentials and identify future priority areas for enhancing their security and stability. The results have represented something of a military–technical revolution throughout the world, albeit at different levels of sophistication.

The strategic focus during the cold war years was on countering the capabilities of the adversary with technological innovations. This also implied denying the adversary as many of the advantages of technology as possible. Centred on major military platforms and weapon systems, the strategy was to constantly improve performance and enlarge the inventory. Although this type of focus continues to be relevant in the context of some regional conflict scenarios, for most militarily advanced nations the focus is now clearly shifting to strategies based on 'systems of systems'. Technological maturity and the compatibility of various systems have made it possible to plan for enhanced military capabilities based on a combination of individual technologies. For instance, one major trend indicates a preference for the integration of long-range, high-precision weapons—which rely heavily on satellite-based reconnaissance and advanced sensors—with the use of fast digital communication links, while another indicates the use of sophisticated airborne or shipboard platforms with customized targeting techniques and a variety of warhead options for intended application objectives.

Another important strategic shift that has occurred is the increased focus on C^3I technologies for conducting integrated war operations with quick reaction times and maximum flexibility. The ongoing revolution in IT has enabled vast arrays of advanced sensors to be used simultaneously for intelligence-gathering systems and for decision-support systems. Compact and fast computers have transformed the battlefield and it is now possible for an individual soldier to possess high situational awareness in real time. With such advanced technological capabilities, older war-fighting doctrines will clearly be replaced by new, tailor-made, flexible strategies that can allow optimal use of military assets under any given circumstances.

Yet another important contribution of technology to future strategic planning is the availability of advanced simulation and war-gaming capabilities.[196] These not only allow major improvements in planning but also help to evaluate the effectiveness of various options for defence strategists and planners. Simulators are also invaluable for high-level training for complex weapon systems. The higher the level of technological sophistication, the higher the demand for comprehensive training, without which high-tech equipment becomes practically useless.

---

[196] See, e.g., the Internet site of the US Department of Defense, Defense Technical Information Center, 'Future warfare: "America's military preparing for tomorrow"', URL <http://www.dtic.mil/jointvision/#>.

Future military strategies will need to take into account some important technology trends. First, the role of dual-use technologies will be far more relevant, with many military capabilities based on civilian technologies. This means that more countries will have access to military capabilities that were available to only a few powers in the past. In a sense this means that the technology gap between the most advanced and the average-level countries will be reduced overall. Hence, strategies and tactics to ensure the best use of available technology will play a larger role in the future.

Given shrinking defence budgets, reduced or changed threat perceptions and acute economic competition, defence producers will tend to change their business practices to make more technological options available to partners and customers in a larger variety of nations or groups. This increases the potential for asymmetric conflict situations and the use of low-intensity warfare techniques. In the regional context, the implications can seriously influence security concerns and military strategies. Export controls and arms control in such situations become increasingly difficult to implement and a sense of lack of control can, in turn, only further fuel the proliferation of conventional weapon technologies. In the regional context, therefore, the trend will be to counter the techno-military capabilities of immediate adversaries in a way that may not be directly related to the technology revolution taking place in the developed world. The regional dynamics of the interplay between strategy and technology will be different in different cases.

Despite the increasing diffusion of technology, in many fields the technology gap will probably remain at the same level or even grow because of the sheer cost and complexity of sophisticated technologies. Stealth technology, smart weapons, ICBMs, strategic cruise missiles and nuclear submarines are examples of technologies that will remain restricted to only a few nations that have the techno-economic means and maturity to possess and use them. Hence, when new technological capabilities are added in the future by a technology leader such as the USA, few other countries may have either the means or the motivation to invest heavily in countering them. The strategies of the target countries may therefore shift towards the acquisition of asymmetric advantages from WMD or to resort to low-technology countermeasures, such as developing assets underground for protection. These are all interesting aspects of technology interplay in the security strategies of the future.

The current military technology transformation is being led by advances in the USA, where the focus is clearly on using IT and space technology for maximum techno-military advantage. The 2003 military operation in Iraq was a convincing demonstration of the new strategy of NCW. It is obvious that for the past decade US planners have redefined their strategic priorities to reflect a mission-oriented strategy rather than the finite goal of fighting two parallel simultaneous wars, as enumerated in military doctrines towards the end of the cold war. The US concept of military transformation envisages the full spectrum of technology enhancement and the introduction of new technologies and capabilities to maximum advantage. Rather than defining an end objective, the strategy appears to be largely evolutionary, allowing for constant change and flexibility. The decision to pull out of the ABM Treaty and to deploy a BMD system clearly indicates the USA's preference for a unilateral approach—with a renewed focus on homeland security—to global issues such as countering terrorism and controlling WMD proliferation. However, even the USA cannot afford to ignore the importance of cooperative security management and multilateral approaches.

In technology terms, new and emerging dimensions of security and threat perceptions must include the security of outer space and space assets, on the one hand, and the real threat to information security, on the other. Protecting information in cyberspace is already proving to be a major challenge. The vulnerability of information-dependent modern societies to information warfare makes this an urgent issue. International norms or formal treaties on these new technology aspects are yet to evolve adequately and the potential for cyber-terrorism remains a real threat to military systems, as well as to civil infrastructures such as financial institutions, power-supply systems and air traffic controls. The ubiquitous nature of cyberspace makes information warfare a potential tool for control as well as a threat. The counter-countermeasure race in IT may not be visible but will certainly spur rapid growth in technological capabilities. The subject of weapons in space, however, is one of high visibility and could transform strategic thinking around the world. If it leads to accentuated insecurity for a larger group of nations, this will be a sad commentary on technological miscalculation at the global level.

The nature of nuclear deterrence has undergone a change because some of the modern advances in conventional weapon technologies have led to such powerful capabilities that the deterrence value of these advanced weapons is substantial. In future the 'system of systems' approach, combining the potential advantages of several high-tech military capabilities, may even provide deterrence comparable to nuclear deterrence, largely because of its international acceptability and the ready usability of these weapons. The international community is already committed to a total ban on CBW. Effective implementation of the BTWC and the CWC could reduce CBW to zero deterrence value. The extreme asymmetric technique of using terrorism as a means to achieve political–military objectives is also close to being universally unacceptable. Hence, there appears to be real potential for an eventual international agreement on universal nuclear disarmament, once there is a paradigm shift in the perceptions of nuclear deterrence.

IV. New patterns of technology diffusion

The major issue that future export control mechanisms will need to address is that of rapid technology diffusion. This is true for several reasons. Technology advances are very fast and spread across a wide spectrum of disciplines. Globalization, as well as unprecedented transparency because of instant worldwide media coverage, has transformed technological awareness all over the world. Even a poor villager in a remote area of a developing country is today more aware of world events and, among other things, of what the richest can afford. It is this awareness that is the most powerful driver for a large part of the world's population to seek technology access and the related opportunities for progress. The have-nots of yesterday did not fully realize what they did not have. In the 21st century such awareness is much more acute and often keeps pace with global developments. In the high-technology sector, industrial practices are constantly changing to remain competitive. In the emerging new technology domain, it is increasingly difficult to define the line between civilian-use technologies and potential dual-use technologies. Innovation is the buzzword and there are several intra-industry information-sharing arrangements across international borders, set up for purely commercial reasons, that defy external controls.

A new dimension of export control problems is the increasing importance of the individual's personal technological knowledge. In sensitive high-technology areas such as nuclear science, propulsion and guidance technologies, simulation techniques, microminiaturization, electronic design and laser technologies, it takes years of first-hand experience to develop expertise. The past five decades have seen a gradual rise of such expertise all over the world—not just in the Western group of supplier countries. These experts are the real repositories of technological knowledge, and they certainly cannot be subjected to typical export control procedures. Similarly, the technological capabilities of a nation depend significantly on its industrial infrastructure and a certain techno-industrial culture that develops over time with techno-economic progress. These are not physical commodities or services that can be controlled through export, unless export controls are made so restrictive as to deny to a country every kind of information on processes and technology. This approach would border on sanction-like measures that are normally used only as punitive actions.

The 21st century situation is thus unlike the 1950s–1990s period, when the majority of technologies were being developed under the umbrella of the military–industrial complexes of the two superpowers, and the situation is continuing to change fairly rapidly. Increasingly, technology is being developed largely by civilian sector enterprises and multinational companies that cut across the globe and work primarily for economic development and commercial benefits.

In the age of globalization, it would be economically inefficient for each nation to seek to develop the whole spectrum of indigenous technology infrastructure. However, given the interplay of sensitive technologies in ever changing international security calculations, it is important for progressive nations to develop core competences in critical and sensitive technology areas. Only countries that have the basic scientific and technological infrastructure and maturity can really absorb high technology and thus benefit from the processes of technology diffusion around the world.

Technological know-how is increasingly being held by private companies that are suppliers to their own governments as well as to others through exports. Apart from military products, ordnance and ammunition, which continue to be controlled largely by governmental agencies, most high-technology components and processes are now dual-use in a reverse mode; that is, it is now civilian technology advances that are creating newer military applications. This process started even before the end of the cold war, and arms control negotiators even then had to decide what to control and how. Almost everything today is dual-use, except for weapon-grade fissile material and some biological precursors that are potential ingredients only for WMD—extreme examples of technologies that are unlikely to be commonplace. If the wide application spectrum of technologies for aeronautics, electronics, propulsion, guidance, sensors or digital electronics is examined, it is difficult to separate out what may be of exclusive military use and therefore a clear candidate for control regimes. These cutting-edge technologies are now held mostly by commercial companies, where technology is more often knowledge-based than defined merely in terms of components and hardware.

A major factor behind technological diffusion is the potential for high technology to command the highest price in the commercial marketplace. Most industrially advanced nations depend heavily on export earnings to remain economically competitive and therefore are subject to forces of competition. These drivers for technology diffusion will always work against the efforts for technology control and export regulation. Regional and global security will demand a delicate balance of these technology-oriented interactions towards protecting security concerns without seriously hampering the course of regional economics and international trade.

In the global economy no company can be competitive without successful exports. Yet, when it comes to the costs for adhering to export controls, it is usually companies rather than governments that pay.

Large companies are acquiring other, smaller companies and mergers are being worked out across the world between partners that were previously rivals. High-tech companies and defence industries are particularly hard-pressed to survive and strongly resent the overbearing export control regulations that restrict their ability to innovate and move ahead in global competition. These companies have to move at real-time speed and cannot tolerate the bureaucratic and legalistic practices of export control regimes. Last but not least, the nature of threats to security and stability have undergone a sea change and most modern technological capabilities are beginning to appear as double-edged swords. Who could have imagined that commercial aircraft could be used to cause such devastation and death as was brought about by the terrorists who carried out the attacks on 11 September 2001?

Another dimension of technological change arises from changes in the way in which technology is inducted into the military domain. In the majority of technologies that have the potential to influence military capabilities, civilian R&D is now in the lead. Earlier it was military technology that was driving civilian industrial development and the military was first to take advantage of new technologies, thereby controlling the civilian adaptation of these technologies at a pace that was acceptable for military priorities. In the 21st century, except for special strategic technologies, civilian R&D is often ahead in most new technology areas. In future, military applications may actually follow after civilian adaptation because induction of technological innovation into military systems is a long process fraught with innumerable and complex considerations of integration, interoperability and cost-effectiveness. Drivers for civilian adaptation are different. Private-sector R&D can no longer afford to be hampered by bureaucratic and security restrictions and is thus racing ahead with faster innovations, better flexibilities and more competitive management infrastructures. This trend will become even sharper in the future and the whole system of technology induction into the military and security apparatus will undergo a sea change. Future technology control regimes will have to adjust quickly to these sweeping changes. Currently, they are incapable of doing so because they were not structured to do anything of the sort.

Controlling exports of emerging dual-use technologies is going to become far more challenging. Decades of technology denial have spurred indigenous R&D in many progressive developing nations and traditional target states such as China and India are becoming important economic and military powers. Several developing nations have emerged as attractive markets for high-technology products. Since these countries are not participants in the 'supply club', there is some concern that the situation may lead to secondary proliferation of sensitive dual-use technologies if exports are opened up, given the futility of denying what already exists. It is interesting to note that, although both China and India have established export control regulations,[197]

---

[197] On India's export control systems see note 81. China announced its establishment of export control systems in 1997 and updated them on 15 Oct. 2002. Anthony, I., 'Supply-side measures', *SIPRI Yearbook 2003* (note 6), pp. 734–35.

Western supplier groups still have difficulty accepting them as partners, whereas some of the former Soviet republics with fledgling economies and doubtful infrastructures for export controls have been welcomed as partners for the future.

The issues, however, are even more complex. These progressive developing nations have become smart buyers that insist on technology transfer with every procurement action. High technology is a buyers' market today, so it is difficult for the supplier to refuse such deals for fear of being beaten by the competition. The effect is that the technology gap between the industrially advanced countries and the developing countries is becoming narrower. Except for the USA, which has relentlessly continued with high-tech R&D and innovation, most other participants in the multilateral export control regimes stand to lose some of their technological edge, and with it the high ground for export controls. The USA, in its turn, may then justifiably look for ways and means to maintain its superiority by exercising unilateral controls against the rest of the world, including some of its former allies, through a unilateral export control regime.

The special feature of modern technology is the high relevance of intangible transfers through the exchange of scientific information among experts. Excessive or intrusive controls such as attempts to control intangible technology transfers through monitoring normal scientific–technical relations among experts should be avoided because they would be counterproductive for the larger goal of wider international cooperation. These are some of the finer nuances of the technology diffusion and technology transfer challenge that must be borne in mind when trying to fine-tune the technology controls of the future.

5. Conclusions

I. In summary

The impact of technology on security has been increasing steadily over the past five decades with impressive advances in modern technologies. The 21st century has begun with many changed parameters relating to technological options and access to and use of technology. This has changed perceptions of what constitutes national security for most nations of the world. The role of technology for defence and development has been enhanced, and trends indicate more significant interplay between technology and matters of safety, security and stability worldwide. Modern technology will inevitably spread to all parts of the world, rich and poor alike, albeit at different levels. However, even moderate levels of technological capability in the wrong hands will have serious security implications. Therefore, the control of potentially dangerous technologies will continue to be an important aspect of ensuring security. Nevertheless, it is not the technology per se that is good or bad. It is the application of technology with dangerous intentions or an irresponsible attitude that is the real cause for concern. If used properly and in a balanced manner, technology is an invaluable key to security, development, progress, cooperation and harmony. The future challenge will therefore be to make the best use of technology to allow universal progress, peace and stability while managing technology interactions and technology advances in such a way as to prevent its careless misuse or dangerous abuse.

As long as nations continue to be unequal in economic wealth and techno-military power, or have differences on ideological, religious or cultural grounds, their relationships will be uneven and problematic. Convergence of views and interests can emerge only in areas where there are common fears or mutual benefit. Progress, peace and stability are major goals shared worldwide. Technology is one common denominator that can help bridge some of the avoidable gaps and can therefore play an important levelling role to enhance international cooperation for long-term benefits for all humanity. There is a far better appreciation today of the effects of technological advances as well as a more widespread political maturity based on many decades of using technology for security needs, economic development, social progress and political leverage. Thus the changing global security scenario presents a unique opportunity to use technology as a binding force to build a safer, progressive and peaceful world society. Reducing the threat of potentially dangerous technologies and weapons as well as controlling the possible misuse of dual-use technologies will remain the major objectives for non-proliferation and export control systems. However, changing patterns of economic progress, global market forces, technology diffusion and new security perceptions will require a radically fresh approach to the effective management of technology in the 21st century, to allow the dynamic balancing of the necessary controls with the imperatives of cooperation and interdependence. The solutions for the future must therefore be packaged imaginatively, and this will require a mature management approach.

While the above can serve as a short conclusion for this report, the issues of technology and security interdependence are numerous and complex. Inter-mixed with political dynamics and human aspirations, the issues of security, stability and peace become even more multifaceted, interdependent and unpredictable. Nevertheless, modern society has managed to progress in an impressive manner by using human creativity and technological tools to its best advantage. Rapid technological advances and concerns over proliferation and misuse have gone hand in hand, and various techniques for control and management have served humanity with varying degrees of success. As we move ahead in this young century, there is a somewhat sombre realization that the challenges of technology management will become even more demanding, while technology continues to provide more choices for better or for worse.

This report highlights this new challenge in the overall context of global security and prosperity, in which technology plays an important role. A special aspect is the inclusion of the demand-side perspective on the issues of technology access, export controls, non-proliferation and disarmament. The aim of the report is to project the viewpoint of a typical progressive developing nation, simply and clearly. If there is perhaps a risk of appearing too critical of Western policies or established international practices, that is not the intention. In fact, the analysis and the arguments in this report are intended to help to evolve future arms control guidelines in a complementary manner. Understanding and accommodation of the other point of view assume a new awareness of the importance of avoiding further alienation of the developing group of nations, in a fast-changing world that is increasingly information-networked and interdependent for economic progress as well as for environmental preservation. External control by itself is not easily acceptable, but controls are necessary in the real world. The success of arms control in the future will depend on the skilful management of controls in order to prevent further alienation among nations and to foster more cooperative interactions.

Observations and inferences

The major observations contained in the various chapters of this report are summarized below to provide a recapitulation of its essence.

Technology issues

1. Technologically advanced nations have secured a higher level of security and stability and face little chance of all-out nuclear or conventional war threatening their security interests.

2. Maintaining technological superiority over other nations, including friendly nations, continues to be an important aspect of safeguarding national security for most sovereign nations.

3. In the new security paradigm of the 21st century, concerns about the proliferation of sensitive technologies such as WMD and missiles have assumed a new relevance because of the rise of non-state players and rogue states that may use such technology for asymmetric advantage.

4. Worldwide communication networks, instant media coverage and market forces within the globally interdependent economies are among the drivers of the modern technology revolution at its interface with everyday life.

5. The enhanced level of awareness about the advantages of advances in technology and the promise of emerging technologies will accelerate the pursuit of high-value technologies throughout the world.

6. The natural process of technology diffusion has created a more interactive and interdependent world and hence an unprecedented

awareness of the need to work together to enhance the prospects for progress and peace.

7. As the costs of developing new technologies rise, nations and companies will have to cooperate in high-technology areas and share those costs to remain at the cutting edge.

8. Politically driven technology denials and discriminatory controls are less likely to succeed in the future. Any non-proliferation gains from such denials or controls will be outweighed by the negative impact of adverse consequences, including international tensions and the loss of a common motive to unite against global threats and problems.

9. It is time to recognize the need to change the old ways of technology control and to initiate processes that can reduce socio-economic polarization and political alienation and foster harmonious cooperation for global security and stability.

10. It will be important to establish an international code for the future use of technology, particularly for exploiting the new options from emerging technologies. This may reduce the challenge of having to control vast arrays of technologies and products with no clear international points of reference.

Proliferation and disarmament

1. Nuclear weapons have proved to have enormous deterrence value and hence will continue to be the strongest currency of power for the foreseeable future.

2. As long as powerful nations continue to depend on nuclear weapons and to enhance such capabilities, other nations will have compelling incentives to acquire similar, competing capabilities.

3. The dangers of WMD proliferation and terrorism are often articulated as major threats to the group of advanced industrial nations. However, they are equally dangerous, if not more so, for the more vulnerable progressive developing nations. Hence, such nations are well motivated to cooperate against proliferation dangers, if the issues can be managed in a fair manner.

4. Although nuclear, biological and chemical weapons are commonly grouped together as WMD, nuclear weapons have the most devastating effects both on people and on military and industrial assets.

5. Chemical and biological weapons target only living organisms and have become universally unacceptable. There is a growing international consensus on eliminating these weapons in the near future.

6. The success of the BTWC and the CWC would be a major achievement that could set the stage for step-by-step universal nuclear disarmament. Conversely, the failure of nuclear disarmament may doom the NPT and seriously harm the success of the BTWC and the CWC.

7. The present US nuclear posture has elevated the potential relevance of nuclear weapons to a new doctrinal level. This not only reduces the threshold for the use of nuclear weapons but also enhances the insecurity of other nations.

8. The compact and efficient conventional weapons of the future could provide effective deterrence. These could be used to help reduce the importance of nuclear weapons.

9. The international responses to all types and cases of proliferation are not uniform. There is increasing recognition that the real concern is about the end use of technology, not the technology itself.

10. Economic and political dimensions are major factors that guide sensitive- and high-technology transfers. Increased globalization and economic competition will accentuate these considerations.

11. As long as WMD exist in some states, the risk of theft or accident will remain. Missile defence technologies will therefore assume increasing importance in the future.

12. Rapid progress will be made in aerospace and missile technologies in many parts of the world. Controlling these technologies will become increasingly difficult and even counterproductive.

13. The future containment of WMD threats will need to focus more on the proliferation or misuse of WMD warheads and their associated technologies than on a select class of delivery vehicles.

14. Intelligence and verification technologies will assume increasing importance in the future, requiring wider cooperation for the effective control of WMD proliferation.

15. Given the interdependent nature of the world, non-proliferation and disarmament objectives must progress hand in hand in order to achieve real and lasting results.

Export controls and technology management

1. Export controls were instituted as a foreign policy tool to deny the enemy the advantage of advanced technology. Now they remain relevant only in the context of controlling proliferation.

2. With the definition of 'enemy' changing dramatically, export controls are gradually becoming convenient tools to maintain technological supcriority—particularly over political opponents.

3. Traditional export controls have succeeded where perceived joint security benefits outweighed any individual gains from non-compliance.

4. Monitoring the activities of and verifying compliance in too many states using a 'catch-all approach' would be an expensive process riddled with technological and political problems. A more cooperative process will be required to ensure future success.

5. The cost of non-compliance is not clearly enunciated in any of the control regimes. Powerful nations follow practices best suited to their economic interests.

6. A technology recipient state regards its actions in seeking advanced technology to enhance its security and economic development as legitimate as the earlier technology acquisitions of now-powerful states.

7. It is imperative for regional and global security that the misuse of technology is comprehensively prevented. However, it is also imperative that a healthy and stable international trade in technology is encouraged to keep pace with overall socio-economic development and continuing advances in technology.

8. The main challenge for the future of non-proliferation and arms control will be to balance the competing interests among the various nations on the issues arising out of technology development, technology trade and concerns regarding technology control.

9. The real focus of 'technology control' must shift to influencing the intentions and decisions of democratic nations or individual dictators who may be inclined to use technology in a manner not conducive to international peace and stability.

10. The future will require a radically fresh approach based on transparent and fair systems that can instil mutual confidence and thus enhance cooperation among the largest possible number of sovereign nations for consolidated action against non-legitimate uses of technology.

II. The future of technology controls

The world in 2004 is very different from that of the 20th century. It has survived four decades of MAD doctrine. Overall quality of life has improved in many countries, with fewer people worried about survival and more people able to pursue their basic aspirations for development and progress. There is a much better understanding of the dangers of nuclear weapons and other WMD technologies. The East–West ideological divide has collapsed, and concepts of cooperative security have proved successful and mutually beneficial. Many military technologies have matured and new levels of transnational awareness have evolved with IT entering every walk of life. This has created a much better appreciation of the potential advantages, as well as the dangers, of advanced technologies. Most importantly, valuable experience has been gained in issues of arms control and disarmament with almost all democratic, progressive nations having agreed 'in principle' on certain international codes of conduct, on the one hand, and the concept of the sovereign status of individual nations, on the other. In terms of dangers to humanity there is little chance of a nuclear disaster of global dimension unless global security is grossly mismanaged in the coming years.

This changed scenario offers a unique opportunity for the world to review its basic policies on non-proliferation and disarmament in order to reduce both the causes of further proliferation and the chances of technology misuse. At the same time, there is enormous potential to enhance international stability through cooperative strategies to collectively counter the real threats of the future so that, eventually, significant disarmament of all WMD by all nations could be achievable in a step-by-step manner—possibly even in the next few decades.

This report focuses on the causes and motivational factors of proliferation, rather than simply presenting a review of the status of non-proliferation or disarmament, which is already well documented.[198] The objective is to analyse trends to identify the 'how' and 'why' of proliferation, particularly in the nuclear and missile fields, and to examine what can be done to curb further proliferation by

---

[198] See, e.g., Anthony (note 197), pp. 727–48; Anthony, I., 'Arms control after the attacks of 11 September 2001', *SIPRI Yearbook 2002* (note 27), pp. 469–88; Anthony, I., 'Multilateral export controls', *SIPRI Yearbook 2002* (note 27), pp. 743–58; and Jones (note 77).

addressing regional security compulsions and other motivational factors. The report aims to support existing non-proliferation efforts by injecting the viewpoint of progressive developing nations that a clear priority should be given to a more universal approach to global economic development, prosperity and security. This cannot be achieved unless global tensions, mutual suspicions and religious fundamentalism are significantly reduced and comprehensively managed. In all this, successful management of technology will play a major role for cooperative progress and collective security.

Given world technology trends and the inevitability of the future diffusion of technologies through the continuing process of globalization, it is imperative that the world learns quickly to manage technology utilization more effectively in order to prevent the future misuse of dangerous technologies. The phenomena of the shrinking global village, instant worldwide media coverage, and the increased transparency and vulnerabilities associated with an emerging IT-dependent world are factors that will greatly influence the future of international technology-management regimes. Extension of the discriminatory policies of the cold war period can no longer succeed. As long as there is open discrimination regarding access to attractive technologies, there will be countries that will aspire to that which is selectively denied. The present climate of forceful unilateralism by the USA has seriously compromised most of the multilateral approaches to non-proliferation and disarmament. There is thus an urgent need to recognize changed priorities and evolve new approaches to non-proliferation and disarmament objectives for the future, on the basis of wider international cooperation among responsible nations for a sharper focus on the real and present dangers to international peace and stability.

The technology controls of the future must be made more focused, more effective and less costly. Initially, export controls were categorized as national security controls and foreign policy controls. Controlling nuclear-missile proliferation was a national security priority and the West's denial of high technology to the Soviet bloc countries was a foreign policy agenda. Over the years the distinction between the two motives has blurred, and the military and security concerns that dominated export control priorities during the cold war no longer have de facto priority over economic considerations. Technology is now being used more for economic competition than for military superiority, and the major technology push now comes from civilian R&D.

Technology control through export controls has served the purpose for which it was designed during the cold war. National controls backed by UN-approved international treaties have performed much better than the closed-door 'supplier group' arrangements such as the MTCR, which have had only limited success. Post-cold war export control systems under various multilateral regimes continue to struggle with a vastly changed environment. There is a tendency to use the technology denial regimes essentially as an instrument of dominance by more powerful nations over less developed or powerful nations, in spite of the legitimate needs of the latter for dual-use technologies. The IT revolution and competitive market forces have created new levels of technology awareness that will continue to grow with time. Technology controls that suppress legitimate technological aspirations create a negative atmosphere for wider international cooperation.

Denial of dual-use technology, by definition, interferes with the right of a sovereign nation to use that technology for its development and security needs, particularly when there is little or no risk of its use as an offensive military capability to create destabilization or anarchy. Nations do not become non-cooperative or 'rogue' overnight. There are patterns of irresponsible behaviour that can and should be addressed through international 'carrot and stick' diplomacy, rather than expecting export control regimes alone to correct the problem. The net benefits of export controls require clearer understanding and articulation in order to balance the benefits of controls against their costs. The total cost implications of export controls for the supplier comprise the implementation costs to governments, the economic costs of the loss of business and political costs linked to the loss of goodwill. Economic interdependence and the effects of globalization will demand increased cooperation among the progressive and democratic nations of the world. The negative implications of excessive export controls will thus assume a new relevance that cannot be ignored.

The probability of full-scale war involving the use of WMD by responsible democratic nations has become remote. Such risks can be further reduced through wider cooperative approaches using the shared values of security and economic development. The real issue for the future is to evolve means of moderating the use of technology, rather than preventing or controlling technology ownership itself by sovereign states. Since the world community stands to gain from the advancement of science and technology, it would be counterproductive to try to control such normal human interactions, although the temptation may be there for those with a 'control fixation'. In the changed scenario of increasing technology diffusion, enhancing security depends more on the identification and careful management of hostile intentions than on containing physical technological capabilities as such. Artificial and unbalanced controls invariably accentuate hostility and exacerbate precisely the forces that need to be controlled.

III. A new approach to technology management

It is in this context that the current use of export controls as a foreign policy tool to suit the changing political priorities of individual nations should now be abandoned. Narrowing the focus of controls, combined with broader international cooperation, should give better results when tackling the real global threats of asymmetric war by terrorists and of WMD proliferation. The focus must now shift to more effective monitoring and verification of compliance at various levels to establish the credibility of the actors involved, both from the demand side and from the supply side. Outright presumption of denial should be reserved only for rogue states and non-state players who should be subjected to the 'catch-all' policies built into the present export control systems. For most other responsible and democratic nations, a universal set of criteria should be evolved for determining in objective fashion (as distinct from subjective 'good guy' and 'bad guy' judgements) what constitutes responsible ownership of technology. A nation's record of past performance in technology management should be used to draw up an international grading system. Access to technology could then be based on a scale linked to a 'Responsible Ownership of Technology' (ROOT) grading. This could provide a much-needed, transparent and fair international system for technology control that would foster confidence and wider cooperation for the future. It could also evolve as the management solution to proliferation and misuse.

The concept of ROOT, introduced here by the author, is a new approach to help fine-tune the technology controls of the future. It would essentially be a system for grading individual states with regard to their proven record of maturity in handling sensitive technology in a responsible manner without compromising international proliferation and security concerns. This should be a 21st century exercise to formulate a management-oriented solution to balancing the problems of dual-use technology proliferation and misuse with the increased need for wider international harmony and cooperation.

ROOT should emerge as a non-political and impartial system for evaluating the 'technological integrity' of a nation by a method that is fully transparent and objective, similar to the system for grading the economic and financial performance of companies and states. The international organization set up to establish ROOT should be a governing council with members drawn from nations that deserve a high ROOT grading, so that these countries (which are also likely to be some of the main technology suppliers) may rise above any concern related to the loss of their controlling status and feel committed to a fair and transparent mechanism for implementing technology and arms control. The ROOT system would have to be evolved carefully after adequate debate and negotiation among nations so that it acquires the international moral authority necessary for such a management body. The onus of control will remain with individual states, but these should take pride in gaining a higher grading in the ROOT system, in the same way that a major reputable company might take pride in its business record in terms of technology dissemination and compliance with export regulations. By holding nations responsible for their intentions and actions, ROOT will help defuse international tensions and foster healthy technology competition among states, on much the same lines as legitimate economic competition.

The authority of the ROOT system should flow out of the technological leverage that the system would wield through its ability to facilitate or deny through the international consensus mechanism built into the system. Over time, its inherent transparency will be in harmony with increasing forces of globalization and the system should therefore remain effective for any future technologies. The existing experience with the UN Register of Conventional Arms, the Australia Group negotiations for dual-use technologies and the HCOC should provide a rich background to assist with developing the planned

ROOT system in a universally just manner that should be acceptable to most nations. The small number of nations that might have reservations about ROOT would reveal themselves during the negotiation process. In that sense, the new criteria for distinguishing between technology recipients could start working right from the beginning of the process.

A possible approach to establishing the ROOT system could start with developing a 25-year historical profile of each nation, based on the available data on technology development, arms production, technology acquisition, and technology and arms transfers. The role of a nation in horizontal transfers as either a supplier or a recipient would be identified and all known data, including intelligence information, would be taken into account in the ROOT analysis.

The country-specific profiles for ROOT grading might consider the following: (a) the political orientation of the country (democratic, autocratic or despotic); (b) the established policies of successive governments towards disarmament and non-proliferation; (c) the consistency and integrity of non-proliferation performance in support of the stated policies; (d) the level of existing export control regulations in force; (e) the existing infrastructure for implementing the regulations; (f) the record of import performance and integrity of end-use certification; (g) the export control performance of the country in the past 15–20 years; (h) the record of indigenous technology development and maturity gained over the past 10–15 years; (i) any evidence of irresponsible behaviour in the past 25 years that created an international crisis or had a major negative impact on international peace; (j) possession of nuclear and missile technologies and whether they have been used as deterrents, offensively or for blackmail; (k) the record of mature handling of advanced technology in terms of accidents and crisis management; (l) the economic stability and economic performance of the country; (m) the human rights history of the country; (n) the environmental management history of the country; and (o) membership of international organizations and the country's contribution to international peace and stability.

This 15-point list offers a representative set of major criteria for ROOT evaluation, where each factor would be graded on a scale of 0–10. For each of these criteria, a variable set of sub-criteria could be evolved for a quantitative assessment of historical records and current performance in specific areas of proliferation concern. A system of negative points must be included for a record of irresponsible export behaviour or a reckless attitude regarding the negative impact of technology leakage. Responsible technological behaviour must emerge as the sole merit for future access to international technology cooperation, while confidence in the system should emerge from a transparent working model and a wide international consensus.

The proposed system of ROOT grading would create unambiguous incentives for responsible behaviour and disincentives against cheating the international system or reneging on international commitments. It would also enhance the perception of fairness in the international system of export controls and technology management.

The ROOT system could also help to realize the other suggestion brought forward in this report regarding 'taller fences around fewer technologies', in a manner that could obtain universal consensus and thereby allow the revised system to focus more on effective implementation. Combined with the 'taller fences' methodology, the new approach would help to trim the list of targets and technologies for control, thus contributing to the reduction of international tensions and creating a harmonious atmosphere on the issues of technology trade and management. This could go a long way towards ensuring lasting international stability and collective security.

It is possible that a new 'international export regulation' could evolve around these recommendations so that it ceases to be merely a control tool and emerges as a cooperation forum for international technology business. As a universally acceptable and transparent system, such a revised system of regulation would carry weight at a lower implementation cost. An appropriate universal UN treaty on 'technology for international peace' could then be evolved for the equitable distribution of technological benefits for legitimate development as well as for the security needs of all sovereign nations. This would undoubtedly provide the long-term route to international peace and stability.

This proposed new approach could sharpen the international focus on technology misuse by using the built-in predictive value of a grading system to identify the responsible and irresponsible use of technology. If established as a fair, just and transparent system it should be acceptable to almost all nations and easy to implement at reasonable cost. It would also be better suited to address future advances in technology and the concomitant future concerns. Such areas of growing concern include emerging advances in militarily useful technologies, military exploitation of outer space, environmental degradation, dwindling natural resources, and increased disease and poverty linked to global demographic imbalances.

A revised international technology-management system must be capable of dealing with all future threats, including future proliferation, economic rivalry, environmental crisis and natural disasters of global dimensions. Technology controls should thus be replaced in the 21st century with technology-management systems that can balance long-term global technology issues with potential security and economic challenges.

IV. Conclusions and recommendations

Threats to global peace and stability have been reduced on two counts. First, the end of the cold war has changed the need for a nuclear balance of terror, and concepts for ensuring national security are being modified to rest on a mix of defensive and offensive capabilities, as tailored to the various regions of the world. Second, there is a wide recognition that the use of WMD, or other weapons capable of mass killing, is unacceptable in modern, progressive, civil society. The international taboo on such weapons, and on senseless killing in general, could go a long way towards reducing threats to security from such weapons and tactics.

However, the potential for regional conflict remains high and the likelihood of limited regional wars and low-intensity conflicts will also continue to be high, particularly where such conflicts involve dictatorial regimes. Promoting democratic governance and enhancing economic development in troubled regions could significantly reduce the chances of military conflict. International intervention will be required to contain many future situations. Non-lethal weapons and techniques as well as sets of economic incentives and disincentives will play an important role in the management of regional conflicts. Future security concerns at the global level will probably be more about economic competition, resource constraints, migration, health and environmental issues. Common security concerns linked to the proliferation of dangerous weapons and the spread of religious fundamentalism and terrorism will require broader consensus and cooperation. The new security environment offers an opportunity for comprehensive cooperation on issues of common concern among responsible, progressive nations.

This report would be incomplete if it failed to highlight the influence of the world technology leader on global security perceptions. As the techno-military superpower, the USA and its policies and priorities will most certainly have a significant impact on the future of technology and security thinking. While the world unhesitatingly supports the USA's counter-terrorism initiative, there are concerns that it is becoming enmeshed with the initiative. Following 11 September 2001 there is a concerted US focus on homeland security that may include the use of new-generation weapons, conventional or nuclear, against the 'bad guys'. Since the demise of the ABM Treaty, there is also a major push to establish a system of missile defence, which may include the weaponization of outer space. These two areas, representing present US priorities, are bound to have major ramifications for future security perceptions and the related technology push–pull effects around the world.

As is often said, peace is not just the absence of war, and security is not just the absence of an imminent threat. Therefore, long-term solutions for peace and security must be based on universally accepted norms that not only are intrinsically fair, but also remain sustainable under various conditions. Such norms and standards alone can create the environment for harmonious relations among a large majority of nations that need to share the benefits of peace and security. In an increasingly interdependent world with national boundaries becoming irrelevant for most international businesses, balancing technology priorities with security perceptions will be crucial for the success of the arms control process in enhancing international security.

The concepts of peace and security have undergone profound changes in the past decade. The spectre of a full-scale nuclear war has receded dramatically and global society has matured enough to allow the elimination of all kinds of WMD, as well as of inhuman methods of war-fighting that often take a severe toll on innocent civilians. Even the poorest sections of society are well aware of the security and prosperity that can be achieved with modern technology. Opportunities for achieving global cooperation on technology utilization for peace and prosperity in the 21st century will thus be attractive and must be seized.

Unfortunately, there are also negative aspects to the changed world scenario. With the counter-balancing forces of the cold war removed, several regional tensions and conflicts have assumed alarming dimensions, and the rise of mercenary forces and religious fundamentalism have fuelled terrorist organizations in many parts of the world. Elements of organized crime and narcotic trafficking have found common identity with the philosophy of terrorism, and the nexus is further aided by irresponsible dictatorial states that are tempted to use such asymmetric forces to settle scores beyond their conventional capabilities. Preventing the misuse of technology for such purposes will require innovative and cooperative approaches.

As long as the potential for regional conflicts remains high, the chances of limited regional wars and low-intensity conflicts will also continue to be high, particularly if such conflicts involve dictatorial regimes. Non-lethal weapons and technologies as well as techno-economic incentives and disincentives must play an important role in the management of future regional conflicts. Promoting democratic governance and enhancing economic development in troubled regions can significantly reduce the chances of military conflict. However, if all else fails, international intervention will be required to contain some future situations. The concepts of pre-emption or punitive action are certainly valuable if used with wisdom and consensus but there is a need for careful articulation of these concepts to gain wide international acceptance, which will add significant deterrence value to such measures.

Future peace and stability concerns at the global level will be increasingly influenced by competition among nations on issues of economic progress, resource constraints and environmental preservation. At the same time, common security concerns linked with proliferation and terrorism should facilitate broader international consensus and cooperation. The changed security environment thus offers a unique opportunity for much broader cooperation on issues of common threat perceptions among the responsible, progressive nations, which not only share common dangers but also must share common dreams.

The future holds out the promise of continuing technology advances and a wide range of new applications and new options in the service of mankind. The impact of the information revolution is already being experienced in everyday life. New technology options from fields such as micro-electromechanical systems, nanotechnology, biotechnology, bionics and artificial intelligence as well as advanced energy-generation concepts could lead to revolutionary changes in the coming decades. Most advances in technology will be essentially for civilian applications and will be generated by civilian research efforts. However, many other technologies, such as those for the exploitation of outer space, will also be prime candidates for military or security-related applications. The concept of military capability is being revolutionized, as witnessed by the impact of IT and NCW on modern war-fighting doctrines.

It is therefore conceivable that the future balance of power in the world will be decided by unique technology capabilities and decisive techno-economic superiority. Killing human beings with WMD will have only terror value, not any real strategic or defence value. The advanced conventional capabilities of smart weapons, as well as the revolutionary new defensive capabilities such as energy weapons, may demonstrate far more strategic value than today's WMD, which cause indiscriminate damage and death. The priorities of technology competition and technology control are therefore likely to change dramatically. For instance, the military use of outer space will add new dimensions to security perceptions and it is quite possible that missile defence technology will graduate to space defence systems in the future. The arms control community must therefore take a fresh look at the definition of dual-use technologies and evaluate the scope and level of controls to achieve an optimal approach to arms control.

The non-proliferation and export control regimes of the past few decades are thus unlikely to provide answers for the future. They have served the purpose for which they were designed in the context of the security perceptions of the cold war years, but 'more of the same' will not do. The new political dimensions of the 21st century and the unprecedentedly fast march of technology will demand a radically fresh approach to technology controls in order that international energies and forces are harmonized towards dealing with common dangers, both intended and unintended.

There are signs of an increasing awareness of the impact of some US policies, pronouncements and actions on non-proliferation and arms control. The USA's withdrawal from the ABM Treaty and the decision to accelerate the establishment of BMD; the continued US investigation of new applications for nuclear weapons and nuclear-weapon technology, together with the revised US nuclear posture; and a greater willingness to supplement multilateral controls with unilateral counterproliferation represent some of the changes to US policy that will have an enormous impact on arms control.

The ABM Treaty, although a bilateral agreement between Russia and the USA, set the international norm against weapons in space and strengthened the objectives of the Outer Space Treaty. The termination of the ABM Treaty could signal a new era of the militarization of space and a new arms race for space defence that may well have a profound effect on air defence and missile defence doctrines for most nations. The spirit of the Outer Space Treaty continues to be relevant, but the treaty needs to be revised and strengthened based on an open, systematic and thorough debate about emerging concerns regarding the militarization of outer space.

The range of technologies for missile defence with space-deployed assets will be of much higher sophistication and capability than what is currently intended to be controlled under the MTCR. The resultant technology dilution will create a peculiar paradox of having to control the low-end technology of ballistic and cruise missiles while nations legitimately aspire to cooperation in the high-end technologies for missile defence. This may well make the MTCR irrelevant and the management of missile, aerospace, surveillance and guidance technologies complex and challenging.

The subtle change in US nuclear posture indicates a movement away from previous commitments on nuclear disarmament and from the negative security assurances given to NNWS and officially included in a 1995 UN Security Council resolution.[199] This will have a profound impact on the NPT regime, and the NNWS signatories may now re-examine their security against nuclear strike. On the other hand, NWS may want to follow the US example of improving their nuclear capabilities and developing space-borne military assets to counter future threats from space. These assets may include high-energy lasers, a new generation of sensors, anti-satellite weapons, hyper-velocity space systems and even the deployment of compact nuclear devices in space for a variety of applications. These and

---

[199] UN Security Council Resolution 984, 11 Apr. 1995, 'on security assurances against the use of nuclear weapons to non-nuclear-weapon States that are Parties to the Treaty on the Non-Proliferation of Nuclear Weapons'. Many states which have forsworn nuclear weapons have requested and received assurances that nuclear weapons would not be used against them. These are known as negative security assurances. Goldblat (note 105), pp. 110–13.

hitherto unknown advances of the 21st century represent the future technologies that will demand radically new approaches to the control and safe management of potentially sensitive technologies.

Making controls more effective

There is a general understanding of what the major international security concerns are, but not a formal universal consensus on their definition or their scope. The political and technological situations today should be used to prioritize security concerns so that revised guidelines for future technology controls can evolve from such an informed appreciation.

Future export controls must be more focused and more effective. It is necessary to narrow their field and sharpen their execution. This can be better achieved by erecting taller fences around fewer technologies. The challenge will be to do so without compromising core international security concerns.

The existing multilateral export control regimes could be replaced by a single universal technology-management regime that would comprehensively address all issues of international technology controls. This would not only increase efficiency but also reduce administrative costs.

Such a universal regime can succeed only if it is perceived as fair and just by all. Hence, a transparent set of criteria for responsible ownership of technology is suggested by the author. The ROOT system is proposed as a new and transparent approach to technology control. Essentially, it would consist of an international system to grade all sovereign nations that want to be technology players in world affairs. The criteria for grading a country could be evolved through international consensus based on a number of parameters and on the country's record of maturity in the handling of technology. The ROOT system would be unique in that it would have built-in qualities of fairness and transparency as well as incentives and disincentives to promote positive behaviour. This proposal deserves a close examination to see whether the existing multilateral export control infrastructure could be fine-tuned in order to move towards such a concept.

The combination of ROOT and the 'taller fences' approach would have significant benefits, not only in terms of enhancing the relevance and effectiveness of technology controls, but also in terms of the suitability of the system to handle future technology advances and challenges.

The new universal technology-management system, if evolved carefully, should supersede all existing multilateral export control regimes to allow technology management for the 21st century to be undertaken without the baggage of 20th century rules, regulations or mindsets.

As emphasized repeatedly, non-proliferation and disarmament must move ahead hand in hand. The first priority should be total disarmament of CBW capacities and comprehensive control of these technologies.

Treaty compliance and verification must receive a high priority. On the one hand, this will require robust cooperation among all responsible nations to make effective use of all of the information available (including from NTM) when monitoring compliance. On the other hand, disincentives for non-compliance must be enhanced.

A minimum international nuclear deterrent may be necessary to counter rogue elements and unknown future eventualities. Such a minimum nuclear capability could be maintained in 'recessed deterrent' mode under an international authority.

A possible move in the direction of nuclear disarmament would be the adoption of a no-first-use doctrine by all nuclear weapon-capable nations. This would reduce the importance of nuclear weapons in strategic thinking and set the stage for step-by-step disarmament leading to the near-total elimination of nuclear weapons. This will be a real challenge for the first decades of the 21st century and for the success of nuclear non-proliferation.

There is an urgent need to generate informed debate on the implications of emerging technologies that may significantly transform security perceptions in the world. This will equip the arms control community with the necessary background to prepare for the future of technology controls.

A new, transparent and fair international security and arms control regime should be considered to oversee the suggested reorganization of international non-proliferation and disarmament activities under the overall ambit of arms control objectives.

In conclusion, this volume highlights the different dimensions of the interplay between technology and security in the changing international scene. The author believes that the future will see an increasingly important role for progressive developing nations such as India, which cannot be left out of the international system for the effective management of security and technology.

The technology push of the 21st century will make it imperative for all nations on the path of progress and peace to unite to tackle the common dangers and reap the maximum benefits from the emerging technologies. Ideally, a unified, fair and transparent international security and arms control regime should meet all such future needs. The real challenge will be to make such ideal solutions work in the non-ideal world. It is essentially a management challenge that will have the best chance of success only if it is supported by the maximum number of partners. Harmony and cooperation among nations must therefore assume a new priority for the future.

This report suggests some new ideas that may be useful for the more effective management of future technologies in the context of preventing undesired proliferation and possible misuse. The ROOT concept in particular merits serious consideration, because it represents a future solution for the management of sensitive technologies. This and some of the other new ideas suggested require further debate, careful examination and considerable work to crystallize the concepts into possible plans for concrete actions.

Index

ABM (Anti-Ballistic Missile) Treaty (1972) 139:
    US withdrawal 88–89, 109, 116, 138
aerospace industry 105
Afghanistan:
    conflict in 1, 34
    Mujahedin 34, 47
    Taliban 47
    terrorism and 30, 47
    USA and 47
    USSR's occupation 47
aircraft, commercial 119–20
aircraft, small manned 86, 87
Algeria 37
Apache helicopters 1
Argentina: nuclear weapons and 50, 76
arms control:
    disarmament and 78
    export controls and 62
    limitations 28
    origins of 59
    purposes of 14–16
    verification 18, 54, 69–70
arms industries:
    export dependence 32
    military expenditure and 4, 31
arms transfers:
    market forces and 31, 103
    markings and 25
    political parameters and 80–81
    pressures 104
    regional conflict and 31, 32
    regional impact of 31–35
    restricting 95
    suppliers 31, 32, 38
    terrorism and 25, 31
    transparency 25
ASAT (anti-satellite) activities 108
ASEAN (Association of South-East Asian Nations) 30
    establishment 30fn.
    members 30fn.
    Regional Forum (ARF) 30
Asia:
    arms imports 39
    conflict in 30, 38
    economic growth 29–30, 38
    proxy wars 30
    tension in 30
Atoms for Peace programme 42–43
Australia Group:
    establishment 13
    members 60–61

ballistic missile defence (BMD):
    case study supporting 70
    deterrence and 69
    development of 88–92
    MTCR and 88–92, 139
    outer space, military use of 17, 138
    see also under names of countries researching
battlefield: enlargement of 6–7
Belarus 72
Biological and Toxin Weapons Convention (BTWC, 1972):
    agreement on 69
    entry into force 12–13
    non-discriminatory 65
    opened for signature 12
    USA and 13
biological weapons: technology simple 52 *see also* CBW
biotechnology 105, 106–107
BLU-114/B 'soft bomb' 2
Boeing 747 aircraft 91, 110
Brazil: nuclear weapons and 50, 76
Brilliant Pebbles 91
Bush, President George W.:
    BMD and 91, 109
    CTBT and 74–75

C^3I 114
Carter, President Jimmy 48

144 TECHNOLOGY AND SECURITY

CBW (chemical and biological weapons):
  banning 54, 117, 141
  consensus about 80
  defence against, research into 54–55
  disarmament, universal 53, 141
  programmes, evidence of 57
  technology simple 52
  unacceptability 65
  verification 52
  see also biological weapons; chemical weapons; weapons of mass destruction
Chechnya 34fn.
chemical weapons:
  technology simple 52
  use of 49, 56
  see also CBW
Chemical Weapons Convention (CWC, 1993):
  agreement on 69
  entry into force 13
  non-discriminatory 65
China:
  BMD 89, 92
  export control 121
  India and 39, 40
  Japan and 103
  missiles 84:
    transfers 85
  MTCR and 85, 87
  nuclear technology transfers 9, 38–39, 74
  nuclear transfers 50
  Pakistan and 38–39, 45, 74
  technology and 103
  USA and 40, 70
  as world power 29
climate change 26
COCOM (Coordinating Committee for Multilateral Export Controls):
  abolition 15
  Cooperation Forum 15
  origin of 14–15, 102
  purpose 61
cold war: technology and 3, 62

cold war end 7:
  export controls and 76
  regional conflicts and 33
  violence and 23
computers:
  battlefield changes and 114
  controls on 6
  improvements 4
  miniaturization 17
  see also IT
Condor II project 82
Conference on Disarmament (CD) 74
conflicts:
  future 35–42
  regional 31, 32, 33, 137
conventional weapons:
  accuracy 25, 33
  lethality 25, 33
  range 33
cruise vehicles, hypersonic 52, 112
CTBT (Comprehensive Nuclear Test-Ban Treaty, 1996):
  negotiations 74
  USA and 73
  value of 53
  verification 74
cyber warfare 105, 116

developing countries:
  alienation of 35, 123
  progressive 79, 95, 121, 123, 125, 129, 142
  technology and 120, 121
directed energy weapons (DEW) 17, 91, 107, 108, 111
drug trafficking 30, 137
dual-use technology:
  controlling 53, 59, 61, 98, 120, 130
  definition 5
  ubiquity of 119
  verification and 70

Egypt 37
Einstein, Albert 70
electronics: miniaturization 3
energy, global 56

energy technologies 105, 107–108
export control regimes:
   companies and 119
   drawbacks 98
   efficacy of 92–110, 140–42
   failures of 14
   gaps in 77, 97
   new approach needed 14
   proliferation and 93
   purpose of 16
   reform needed 16, 18, 82, 99–100, 120, 130
   supplier arrangements 92
   'taller fences around fewer technologies' 134, 140
   use and 98, 99
   verification 98
   viability and 7
   see also following entry, names of regimes and technology control
export controls:
   arms control and 62
   China and 81
   consensus on 7
   cost of 104, 130, 134
   demand side and 77, 78
   economic competition and 79
   economic growth and 7
   environment changing 76–82
   intent and 99
   IT and 104
   non-proliferation 80, 127
   political motivation 77
   priorities for changed 79
   purpose of 127
   rationale for changed 78
   reform needed 82, 140
   strictness unreasonable 99
   supply side 78
   targets of 81
   technology diffusion and 7
   technology management and 127
   terrorists and 78
   see also preceding entry and technology control

F-16 aircraft 81
fissile materials cut-off treaty (FMCT) 53, 72, 73:
   India and 75
   Israel and 75
   Pakistan and 75
fissile materials transfer: economic and political dimensions 52
France:
   arms exports 38
   BMD 89
   nuclear technology transfers 9

Gandhi, Prime Minister Rajiv 74
Geneva Protocol (1925) 59
Germany:
   nuclear technology transfers 46
   nuclear weapons and 51, 75
Ghauri II missile 85
Global Positioning System (GPS) 1, 2, 84
globalization: technology diffusion and 4, 103, 117
Group of Seven (G7): MTCR and 10–11, 83
Gulf Cooperation Council (GCC) 89–90
Gulf War (1991) 57, 76, 97

Hague Code of Conduct (HCOC, 2002) 12, 88, 132
Hague Conventions (1899 and 1907) 59
HEU agreements, Russia–USA 75
hydrogen fuel 107

IAEA (International Atomic Energy Agency):
   establishment of 71
   Iran and 50
   North Korea and 48
   purpose of 8, 59–61
India:
   BMD 89
   China and 38, 40
   CTBT and 73
   economy 29, 39
   export control 121

missiles 12, 84
National Security Advisory Board 45
nuclear disarmament and 74
nuclear tests 45
nuclear weapons 9, 43, 44–45, 51:
    second-strike capability 46
Pakistan and 38–40, 46
PTBT and 73
USSR and 38
Internet 4
Iran:
    IAEA and 50
    missiles 12:
        transfers to 85
    nuclear ambitions 37, 50, 56, 75
    nuclear technology transfers to 50
    nuclear transfers to 50
    WMD capabilities 79, 97
Iran–Iraq War (1980–88) 49, 77
Iraq:
    CBW stockpiled 76
    chemical weapons use 49
    missiles 12
    nuclear weapons and 37, 76
    UN inspections 13
    USA and 97
    WMD capabilities 79, 97
Iraq war (2003):
    advanced weapons in 1–2, 116
    air defence 1
    C^3I and 1
    computers 2
    satellites 1
    security elsewhere and 28
    technology and 1, 2
    UN and 2–3
Islamic extremism 36, 46
Israel:
    BMD 89
    missiles 12, 84
    nuclear ambiguity 43
    nuclear technology transfers to 9, 73
    nuclear weapons 9, 37, 43–44:
        second-strike capability 44
    USA and 36, 37, 44

IT (information technology):
    battlefield revolution 17
    biotechnology 106
    effects of 105–106, 114, 130, 138
    importance of 6
    verification and 106
    warfare and 116
Italy: BMD 89

Japan:
    China and 103
    nuclear weapons and 51, 75
    technology player 29
Joint Direct Attack Munition (JDAM) 2

Kashmir 39, 46
Kazakhstan 72
Khatami, President Mohammad 50
Kim Jong-il, President 40
Korea, North:
    China and 40
    economy 40
    IAEA and 48
    isolation 40
    migration from 41
    missiles and 41, 48
    NPT withdrawal 48, 75
    nuclear weapons and 41, 48–49, 75
    Pakistan and 48
    plutonium 48
    reactor de-fuelling 48
    Russia and 40
    South Korea and 40–42
    USA and 41
    WMD and 40:
        capabilities 79, 97
Korea, South:
    missiles 85
    North Korea and 40–42
    nuclear weapons 51, 75

lasers 90, 91, 110, 111
Libya:
    nuclear ambitions 37
    UK–USA deal with 37

INDEX 147

man-portable air defence systems (MANPADS) 33–34
market forces 5
media 117, 129
Middle East:
    arms imports 38
    CBW capabilities 37
    conflict in 36–38
    insecurity in 37
    Islamic extremism 36
    terrorism in 36
    wealth 36–37
    WMD and 36, 37
military aid 32
military expenditure 4, 31, 115
missile technology:
    commercial pressures 83
    control of 83
    dual-use nature 86
    evolution of 10
    space-launch technology and 12, 87–88
    trade in 41
Missile Technology Control Regime (MTCR):
    BMD and 88–92
    Category I missiles 11–12
    Category II missiles 12
    commercial pressures 87
    cruise missiles and 85
    failures 87–88, 91
    limits of 88
    members 60–61, 83, 87
    modification 28, 57, 84
    origin of 10
    purpose of 10–11
    sanctions absent 86
    successes 87
    as supply cartel 11
    technology focus 86
    UAVs and 85
    undercutting rivals 86
    user focus 84, 85, 86
missiles:
    developing countries acquiring 82–83
    nations' capabilities 12

    non-state actors and 84
    proliferation 28, 83
    terrorists and 84
    transfers 83
missiles, cruise:
    guidance 84
    technology for 34, 84–85
multinational companies 5, 79
Musharraf, President General Pervez 47
mutual assured destruction (MAD) 8, 16, 22

nanotechnology 107
NATO (North Atlantic Treaty Organization): enlargement 15
Netherlands 46
network-centric warfare (NCW) 1, 17, 116, 138
No Dong missile 85
non-lethal weapons 135, 137
non-proliferation:
    focus change 54–55
    future challenges 55–58
    see also proliferation
non-state actors 33–34, 57, 68, 84, 124, 131
NPT (Non-Proliferation Treaty, 1968):
    challenges 55–56
    disarmament commitments 8, 71, 72:
        ignored 80
    entry into force 71
    failure of 71
    focus of 72
    indefinite extension of 9
    late accessions 43
    new NWS and 43, 71
    review conferences 53, 71, 72, 75
    status 75
    US nuclear posture change and 139
nuclear deterrence 66–70, 117, 125
nuclear disarmament, universal 18, 27, 51, 53, 55, 56, 72, 141
nuclear energy 56
nuclear proliferation 8–9:
    early stages of 43
    preventing 59

regional motives 69
vertical 75
Nuclear Suppliers Group (NSG):
  members 60–61
  purpose of 8
nuclear technology transfers 8, 9, 38–39, 46, 50, 73, 74
nuclear weapon states: status and 8, 42
nuclear weapons:
  accuracy 27
  aspirants to 27, 50
  basic device 24
  CBW and 68–69
  coercion, resisting 66
  costs of 63–64
  dependence on 3, 18
  doctrines 68–69
  efficiency 27
  large stockpiles of 64, 65
  new uses for 3
  no-first-use 55, 56, 68–69, 141
  numbers 8
  proliferation of 8–9, 63–70:
    control of 65
  regional security and 50
  size reduction 27
  status and 8
  use unacceptable 64
  weak regimes and 54, 68
  weight reduction 27

oil 36, 113
Organisation for the Prohibition of Chemical Weapons (OPCW) 61
outer space:
  BMD and 91, 92
  military use of 17, 108, 138
  peaceful use of 88, 112
  security 116
  weaponization of 69, 108, 116
Outer Space Treaty (1967) 108, 109, 139

PAC (Patriot Advanced Capability) system 90, 109
Pakistan:
  Afghanistan and 47
  arms imports 39
  China and 38–39, 73, 74
  India and 38–40, 46
  Islamic extremism 46
  military aid to 38, 39
  missile transfers to 85
  missiles 12
  North Korea and 49
  nuclear technology transfers to 46, 74
  nuclear tests 47, 73
  nuclear transfers to 9, 38–39
  nuclear weapons 9, 43, 46–47, 74
  terrorism and 30, 39, 46, 81
  USA and 38, 39, 46, 47
Palestinians: missiles 12
Partial Test Ban Treaty (PTBT, 1963) 71, 73
proliferation:
  causes of 42–51
  disarmament and 125–26
  international reactions to 52
  non-nuclear 57
  prioritizing concerns 51–55
  threat seriousness 42
  see also non-proliferation; nuclear proliferation
Pugwash Conferences on Science and World Affairs 70

al-Qaeda 39, 81
Qadhafi, President Muammar 37

RAND 3, 9
Reagan, President Ronald 17, 88
religious fundamentalism:
  causes of 23
  spread of 17, 18, 135, 137
  WMD and 54
resources: constraints on 29
'Responsible Ownership of Technology' (ROOT) 131–35, 140
rich and poor, gap between 26
rogue states:
  export controls and 81, 131
  terrorists and 78
  threats from 23, 54

WMD and 65, 67, 141
Russell, Bertrand 70
Russia:
    arms exports 31, 38
    BMD 89, 92
    economic recovery 29
    MTCR and 85, 87
    nuclear transfers 50
    see also Union of Soviet Socialist Republics

Saddam Hussein, President 76
satellites:
    BMD and 108, 111
    damage to 111
    early warning 108
    imagery improved 1
    military use 17
    missile guidance 2
Saudi Arabia 37
security:
    changed perceptions of 122, 135, 136
    economic competition and 29
    factors affecting 36
    global:
        changing 123, 135
        USA's effect on 136
        USA's perceptions and 27
    regional 22
    threats to 22–23
Semiconductor Manufacturing International Corporation (SMIC) 104
sensors 114, 139
Shaheen missile 85
smart weapons 2, 138
South Africa:
    missiles 84
    nuclear weapons and 50, 75–76
space debris 108, 111
space launch technology 10, 52
space technologies 105:
    diffusion of 108–13
Sputnik launch 10, 62
Strategic Defense Initiative (SDI) 17, 88
strategic weapons, non-nuclear 18
strategic weapons, nuclear 17

Sweden 80
Syria 37
systems technology 6
system of systems approach 117

Taiwan 70, 81
technology:
    advantages 122–23
    civilian development 118–19, 120, 130
    civilian–military interplay 5, 117
    combination of individual 114
    costs 125
    defensive capability 67
    denial regimes 104, 130
    development costs 5
    economic gain and 67
    espionage and 62, 130
    future trends 113–17
    impact of 104–13
    individuals' expertise and 118
    international code for 125
    issues discussed 124–25
    military 5
    military importance of 3
    security and 1, 14
    see also following entry and dual-use technology
technology control:
    arms control policies and 59
    challenges 123
    demand side 78
    end user 9
    future of 128–31
    management-oriented approach 78, 140, 141
    new approach to 131–35
    new challenges 4
    reform needed 123, 125, 129, 131–35
    security and 122
    supply side 78
    technology focused on 9
    see also export controls
technology control regimes: scope flawed 13–14

technology diffusion:
  conflict and 36
  control of 97, 117
  definition 103
  globalization and 4, 104, 117
  inevitability of 5
  'intangible' 103, 121
  military expenditure and 4
  new patterns 117–21
  non-proliferation 7
  see also following entry
technology transfers:
  conventional weapons 9
  demand for 121
  globalization 78–79
  market forces and 5, 130
  see also preceding entry and nuclear technology transfers
terrorism:
  countering 23, 25
  spread of 18, 135
  state support 4, 24
  support structures 23
  unpredictability of 24
  war on 25–26, 31, 57
  see also following entry
terrorists:
  CBW and 67
  concealment and 81–82
  export controls and 81
  intelligence on 24–25
  technology used by 4
  threats from 23
  WMD and 24, 54, 65, 67
  see also preceding entry
Theater High-Altitude Area Defense (THAAD) 90
thermonuclear fusion 107
threat perceptions:
  factors influencing 22–23
  future 35–42
  regional 22
threats: asymmetric 23, 33, 53, 67
Tomahawk missile 2, 34
TRW Instruments 110

Ukraine 72, 85
Union of Soviet Socialist Republics (USSR):
  BMD 88
  collapse of 56
  missile transfers 83
  nuclear weapons 63, 82, 102–103
  scientists 103
  technological status 102
  technology, economic burden of 102
  USA and 46
  see also Russia
United Nations:
  arms transfers and 32–33
  Charter 66
  degradation of 28, 35, 82
  Disarmament Commission 32
  disarmament, special session on 74
  effectiveness diminished 30
  marginalization of 35, 82
  P5 27, 71, 80, 81
  Register of Conventional Arms (UNROCA) 32–33, 95, 132
  Security Council Resolution 984 139
United States of America:
  aid given by 32, 33
  arms control, criticisms of 78
  arms exports 31, 33, 38, 81
  Atoms for Peace programme 42–43
  BMD 88, 89, 90–92, 108–111, 116, 138
  Congress 74
  DARPA 3, 102
  export controls and 97
  Foreign Assistance Act (1961) 46
  Glenn–Symington Amendment (1977) 46, 47
  homeland security 25, 116
  international policing 35
  less powerful nations and 29
  military superiority 1
  military technology 1
  missile threats against 89
  missile transfers 83

National Strategy to Combat
  Weapons of Mass Destruction 27,
  28
nuclear posture 138, 139
Nuclear Posture Review (2001) 3
nuclear technology transfers 9, 46
nuclear weapons 63, 82, 102–103,
  126, 138
Pressler Amendment (1985) 47
responsibility of 57
security assurances 139
technology:
  advances 57, 102, 116
  protection of 102
  as weapon 102
terrorist attacks on (2001) 4, 23, 84,
  87, 120
unilateralism 30, 116, 121, 129
weather modification 112–13
unmanned air vehicles (UAVs):
  improvements in 1
  poor man's use of 35
  technology behind 34
  WMD delivery and 57

verification technologies 54, 69–70
very large-scale integration (VLSI) 3

warhead technology 52, 54
Wassenaar Arrangement (WA):
  demand side 96
  establishment 93
  evolution of 94
  focus 96
  founders 33
  looseness of 16
  members 60–61, 94
  motives behind 93, 95
  negotiations founding 94
  origins of 15, 93
  political considerations 96
  purpose of 61
  supply side 96
weapons:
  accuracy increased 18
  miniaturization 17

see also following entry and
  conventional weapons; nuclear
  weapons
weapons of mass destruction (WMD):
  accident and 126
  controlling, consensus about 80, 135
  conventional weapons and 3
  theft of 24, 126
weapons of mass disruption 6–7
weather 112–13
World Bank: effectiveness diminished
  30
world order, changing 22–31
World War II 3, 63, 64

Yemen 1

Zangger Committee, members 60–61